SpringerBriefs in Computer Science

T0183124

More information about this series at http://www.springer.com/series/10028

Quang-Dung Ho • Yue Gao
Gowdemy Rajalingham • Tho Le-Ngoc

Wireless Communications Networks for the Smart Grid

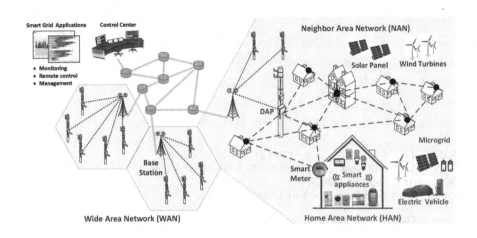

 Springer

Quang-Dung Ho
Department of Electrical
   and Computer Engineering
McGill University
Montreal, QC, Canada

Gowdemy Rajalingham
Department of Electrical
   and Computer Engineering
McGill University
Montreal, QC, Canada

Yue Gao
Department of Electrical
   and Computer Engineering
McGill University
Montreal, QC, Canada

Tho Le-Ngoc
Department of Electrical
   and Computer Engineering
McGill University
Montreal, QC, Canada

ISSN 2191-5768　　　　　　ISSN 2191-5776 (electronic)
ISBN 978-3-319-10346-4　　ISBN 978-3-319-10347-1 (eBook)
DOI 10.1007/978-3-319-10347-1
Springer Cham Heidelberg New York Dordrecht London

Library of Congress Control Number: 2014948258

Printed on acid-free paper

Springer is part of Springer Science+Business Media (www.springer.com)

Preface

In order to enhance the efficiency and reliability of the power grid, diversify energy resources, improve power security, and reduce greenhouse gas emission, many countries have been putting great efforts in designing and constructing their smart grid (SG) infrastructures. Smart grid communications network (SGCN) is one of the key enabling technologies of the SG. However, a successful implementation of an efficient and cost-effective SGCN is a challenging task.

This Springer brief gives a comprehensive overview of SGCN by investigating its network architecture, communications standards, and quality-of-service (QoS) requirements. Promising wireless communications technologies that could be used for the implementation of the SGCN are also addressed. In addition, two candidate protocols for the neighbor area network (NAN) segment of the SGCN are investigated and compared in order to identify their strengths, weaknesses, and feasibilities. Especially, a proactive parent switching mechanism to improve the resilience of NANs against smart meter failures is also presented and evaluated. As an attempt to identify possible future research trends, this brief also outlines a number of technical challenges and corresponding work directions in the SGCN.

The target audience of this informative and practical brief is researchers and professionals working in the field of wireless communications and networking. The content is also valuable for advanced-level students interested in architecture design, routing protocol development, and implementation of wireless mesh networks.

We would like to acknowledge the financial supports from the Natural Sciences and Engineering Research Council of Canada (NSERC) through a NSERC discovery grant and the NSERC Smart Microgrid Network (NSMG-Net) and the

Fonds Québécois de la Recherche sur la Nature et les Technologies (FQRNT) via a scholarship.

Finally, we dedicate this work to our families.

Montreal, Canada

Quang-Dung Ho
Yue Gao
Gowdemy Rajalingham
Tho Le-Ngoc

Contents

Acronyms

1G, 2G, 3G, 4G	First, Second, Third, Fourth Generation
3GPP	3rd Generation Partnership Project
6LoWPAN	IPv6 over Low-Power Wireless Personal Area Network
ACK	Acknowledgment
ADA	Advanced Distribution Automation
AMC	Adaptive Modulation and Coding
AMI	Advanced Metering Infrastructure
ANSI	American National Standards Institute
AP	Access Point
API	Application Programming Interface
APP	Application
AV	Autonomous Vehicle
BAN	Building Area Network
BLE	Bluetooth Low Energy
BT	Bluetooth
CAPEX	Capital Expenditures
CCHP	Combined Cooling, Heat, and Power
CDF	Cumulative Distribution Function
CDMA	Code-Division Multiple Access
CSL	Coordinated Sampled Listening
CSMA-CA	Carrier Sense Multiple Access with Collision Avoidance
CSP	Concentrated Solar Power
CT-IAP	Communications Technology Interoperability Architectural Perspective
CTS	Clear to Send
DA	Distribution Automation
DAG	Directed Acyclic Graph
DAP	Data Aggregation Point
DDoS	Distributed Denial-of-Service
DER	Distributed Energy Resource
DG	Distributed Generation
DIFS	Distributed Coordination Function Interframe Spacing

DIO	DODAG Information Object
DIS	DODAG Information Solicitation
DODAG	Destination-Oriented Directed Acyclic Graph
DR	Demand Response
DS	Distributed Storage
DSL	Digital Subscriber Line
DSSS	Direct Sequence Spread Spectrum
EDR	Enhanced Data Rate
EIFS	Extended Interframe Spacing
EMI	Electromagnetic Interference
EPS	Electric Power System
ETSI	European Telecommunications Standards Institute
ETX	Expected Transmission Count
FAN	Field Area Network
FDMA	Frequency-Division Multiple Access
FHSS	Frequency Hopping Spread Spectrum
FLIR	Fault Location, Isolation, and Restoration
GEO	Geographic Routing
GF	Greedy Forwarding
GPSR	Greedy Perimeter Stateless Routing
HAN	Home Area Network
HART	Highway Addressable Remote Transducer
HEMS	Home Energy Management System
HSDPA	High-Speed Downlink Packet Access
IAN	Industrial Area Network
IAP	Interoperability Architectural Perspective
IC	Internal Combustion
ICT	Information and Communications Technology
IEC	International Electrotechnical Commission
IED	Intelligent Electronic Device
IEEE	Institute of Electrical and Electronics Engineers
IETF	Internet Engineering Task Force
IP	Internet Protocol
IPv4	Internet Protocol version 4
IPv6	Internet Protocol version 6
ISM	Industrial, Scientific, and Medical
ISO	International Organization for Standardization
IT-IAP	Information Technology Interoperability Architectural Perspective
ITU	International Telecommunication Union
IoT	Internet of Things
LAN	Local Area Network
LR-WPAN	Low-Rate Wireless Personal Area Network
LTE	Long-Term Evolution
LTE-A	LTE-Advanced
M2M	Machine-to-Machine
MAC	Medium Access Control
MAN	Metropolitan Area Network

MFR	Most Forwarding Progress within Radius
MIMO	Multiple-Input Multiple-Output
MP2P	Multi-Point-to-Point
MPDU	MAC Protocol Data Unit
MTC	Machine-Type Communications
MTX	Mean Transmission Time
NAN	Neighbor Area Network
NFC	Near-Field Communication
NFP	Nearest with Forwarding Progress
NIST	National Institute of Standards and Technology
O-QPSK	Offset Quadrature Phase-Shift Keying
OF	Objective Function
OFDM	Orthogonal Frequency-Division Multiplexing
OFDMA	Orthogonal Frequency-Division Multiple Access
OMS	Outage Management System
OPEX	Operational Expenditure
OSI	Open Systems Interconnection
P2MP	Point-to-Multi-Point
P2P	Point-to-Point
PAP	Priority Action Plan
PDR	Packet Delivery Ratio
PEV	Plug-in Electric Vehicle
PHY	Physical Layer
PIFS	Point Coordination Function Interframe Spacing
PLC	Power Line Communications
PMU	Phasor Measurement Unit
PPS	Proactive Parent Switching
PS-IAP	Power Systems Interoperability Architectural Perspective
PV	Photovoltaic
QoS	Quality-of-Service
RF	Radio Frequency
RFID	Radio-Frequency Identification
RNC	Radio Network Controller
ROLL	Routing over Low-Power and Lossy Networks
RPL	Routing Protocol for Low-Power and Lossy Networks
RTS	Request to Send
SAE	System Architecture Evolution
SAS	Substation Automation System
SC-FDMA	Single-Carrier Frequency-Division Multiple Access
SCADA	Supervisory Control and Data Acquisition
SDN	Software-Defined Networking
SDO	Standards Developing Organization
SEP	Smart Energy Profile
SG	Smart Grid
SGCN	Smart Grid Communications Network
SGIP	Smart Grid Interoperability Panel
SGIRM	Smart Grid Interoperability Reference Model

SIFS	Short Interframe Spacing
SIG	Special Interest Group
SM	Smart Meter
ST	Slot Time
SUN	Smart Utility Network
TCP	Transmission Control Protocol
TDMA	Time-Division Multiple Access
TOU	Time-of-Use
UDP	User Datagram Protocol
V2G	Vehicle-to-Grid
VAR	Volt-Ampere Reactive
VM	Virtual Machine
VoIP	Voice over Internet Protocol
WAN	Wide Area Network
WASA	Wide Area Situational Awareness
WMN	Wireless Mesh Network
WPAN	Wireless Personal Area Network
WSAN	Wireless Sensor and Actuator Network
WSN	Wireless Sensor Network
WiMAX	Worldwide Interoperability for Microwave Access

Chapter 1
Introduction

1.1 Today's Power Grid

The power grid (also referred to as the electrical grid) is an interconnected network that generates and delivers electric energy to its consumers. In the early days of electricity (the late 1800s), energy systems were small and localized. As technologies evolved, larger plants and longer transmission and distribution lines were constructed to provide more electricity to a growing population. As a result, today, the power grid has become extraordinarily complex and widespread. The U.S. electric, for example, serves more than 143 million residential, commercial, and industrial customers through more than 6 million miles of transmission and distribution lines owned by more than 3,000 highly diverse investor-owned, government-owned, and cooperative enterprises [1, 2].

The structure and constituents of today's power grid are shown in Fig. 1.1. Specifically, it consists of electrical power generation systems, power transmission and distribution grids, substations and consumers. The grid starts at the power plants that include the facilities for generating bulk electrical energy. Conventional power generators mainly run on non-renewable energy sources including nuclear, coal, natural gas, and petroleum. They can also run on hydro or geothermal energy. These power plants are quite large in order to take advantage of the economies of scale and are often located near coal mines, oil rigs, dam sites, etc., that are far away from heavily populated areas. The generated electrical energy is stepped up to higher voltages by generation substations and then connected to the transmission grid. In order to reduce line losses, typical voltages for long distance transmission are in the range of 138–765 kV. High-voltage electric power cables of this grid deliver the bulk power to distribution facilities near populated areas. The transmission grid covers a vast area (e.g., national or across international boundaries). On arrival at distribution substations, the power is stepped down from transmission-level voltages to distribution-level voltages and then enters the distribution grid. Distribution

Q.-D. Ho et al., *Wireless Communications Networks for the Smart Grid*,
SpringerBriefs in Computer Science, DOI 10.1007/978-3-319-10347-1_1

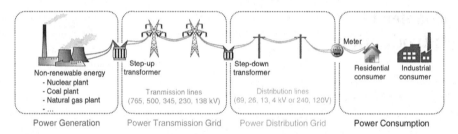

Fig. 1.1 Overall structure, functional blocks and elements of today's power grid

voltages are typically less than 10 kV. Finally, at the service location, the power is stepped down again from the distribution-level voltages to required service voltages and delivered to consumers. The power consumer refers to either a residential, industrial or commercial entity that consumes power from the distribution grid.

The existing power grid exhibits the following key attributes. First, power is generated in large centralized plants using mostly non-renewable energy sources. Second, the energy uni-directionally flows from the plants to consumption sites. Third, most power equipment have yet to be computerized and automated. Real-time monitoring and control is mainly limited to a small fraction of generation and transmission devices. Fourth, the configuration, operation and management of the grid are not flexible and thus it is impossible to either inject electricity from alternative sources at any point along the grid or to enable new services desired by the consumers. And finally, consumers do not participate in the operations of the grid to optimize their power usage. They simply receive monthly bills informing them of their power consumptions and costs.

1.2 Drivers and Objectives of the Smart Grid

Even though the power grid has made vital contributions to the growth of the national economy, industry and the quality of daily life, its physical elements as well as fundamental operation, maintenance and management methods have not been innovated in the past century. It has remained essentially unaltered while the number of consumers and their demands has grown exponentially. In many countries, the power grid supplies electricity to hundreds of millions of households, businesses and industries. At the residential level, power consumption has been increasing with the introduction and widespread use of air conditioning and heating, water heating, refrigerators, dishwashers, televisions, washers, dryers, computers, etc. Particularly, with rapidly advancing technologies and the availability of various national and provincial incentive programs, plug-in electric vehicles (PEVs) are becoming increasingly more affordable to consumers. Their higher rate of penetration is going to add significant electricity consumption. The study in [3] estimates that PEVs increase residential customer load by 33–37 % in the next decade. Further, the government has recognized the vital role of the national power grid to homeland security and public safety. However, lack of investment, combined with aging equipment, has resulted in an inefficient and increasingly unstable electric system [4]. Obviously, older equipment has higher failure rates and thus leads to higher power interruption rates that affect both the economy and society. Older assets and facilities also result in poorer power quality and require higher costs for inspection, repair and restoration. Aging transmission and distribution grids, with a low level of automation and control, lead to significant power loss as well slow and inefficient responses to system failures or faults. Additionally, the majority of devices used to generate and deliver electricity have not been automated and computerized. To that effect, over the last century, utility companies have had to send workers to the fields to gather data necessary for operating, maintaining and trouble-shouting the power grid. For example, the workers read meters, search for failed power lines or broken equipment and measure power quality.

Aside from the aging power system, the lack of sufficient facilities and crisis handling procedures for fault detection, localization, isolation and service restoration is another critical issue in today's power grid. Accidental break-downs or natural disasters could potentially have a devastating impact on the supply, generation and delivery of power to large areas of the country. For example, lightning or untrimmed trees may cause short-circuit faults on transmission and distribution lines; tornados, storms and earthquakes may destroy power plants, electricity poles and substations. In fact, many incidents causing cascading failures and extended blackouts over vast areas have been observed. These incidents not only hurt the economy but also pose critical threads to security and safety at national and international scales. The Northeast blackout in 2003 is an illustrative example. It was a widespread power outage that occurred throughout parts of the Northeast and Midwest United States and the Canadian province of Ontario on Thursday, August 14, 2003, just before 4:10 p.m. Eastern Daylight Time [5]. While some areas had their power restored by 11 p.m., many did not get power back until 2 days later [6]. The blackout affected an estimated 45 million people in 8 U.S. states and 10 million people in Ontario. Investigations revealed that the blackout's primary causes were a software bug in the alarm system at a control room and the lack of effective real-time diagnostic support. The grid operators were unaware of the need to re-distribute power after overloaded transmission lines touched an overgrown tree limb and short-circuited [7].

Furthermore, over the last decades, there has been increasing concerns on the scarcity of non-renewable energy resources and environmental pollution. Fossil fuels such as coal, natural gas and crude oil, that take hundreds or even millions of years to form naturally, have been excessively used by conventional large power plants. These resources are finite and if their current worldwide consumption rates continue, they could be gone in 35–40 years [8]. For uranium, once mined, it can never be replaced. In addition to scarcity, when fossil fuels are burnt, they release a large amount of toxic substances (i.e., carbon monoxide, nitrogen oxides, hydrocarbons and sulfur oxides) into the atmosphere that lead to numerous human diseases and air pollution. Currently, it is estimated that electric power generation represents approximately 25 % of global greenhouse gas emissions that significantly contribute to climate change [4,9]. Further, extracting and transporting oil have also led to oil spills and slicks that pollute water resources and damage surrounding natural environments. As for nuclear power, it has raised big safety and environment pollution concerns. Notably, nuclear and radiation accidents result in significant death tolls, fetal diseases and vast century-long uninhabitable areas. Chernobyl (April 26, 1986, Ukraine) and Fukushima Daiichi (March 11, 2011, Japan) are the two most catastrophic nuclear accidents. They are classified as level-7 events on the international nuclear event scale. These events lead to massive death tolls from not only the event itself but also its long-term effects. Additionally, hundreds of thousands of residences in areas surrounding the contamination have to be evacuated. Evacuees exposed to radiation and generations of their descendant suffer from a higher rate of cancer development. Massive amount of radioactive soil and water is another critical problem that costs hundreds of billions of dollars and several decades or even centuries to clean up. Therefore, in order to prevent such catastrophes in the future, many countries, e.g., Germany, have decided to shutdown and decommission their nuclear power plants [10].

These issues and concerns have urged many nations to enhance the efficiency and reliability of their existing power grids as well as to seek alternative sources of renewable, reliable and high-quality electricity. This drives the need for the smart grid (SG). By definition, the SG "is an automated, widely distributed energy delivery network characterized by a two-way flow of electricity and information, capable of monitoring and responding to changes in everything from power plants to customer preferences to individual appliances" [11]. In a nutshell, the SG means a "modernized" and "computerized" electric utility grid. The primary objectives of the SG are to allow utilities to generate and distribute electric power with enhanced quality, efficiency and reliability, to reduce the contribution of the power grid to climate change and to allow consumers to optimize their energy consumption.

Firstly, in order to reduce power transmission loss and environmental pollution, extensive use of distributed energy resources (DER) and renewable energy are important features of the SG. DER encompasses a wide range of small-scale distributed generation (DG) and distributed storage (DS) technologies located close to where electricity is used (e.g., a home or business). They can operate either in parallel with or separately from the utility distribution system. DG technologies are internal combustion (IC) engines, gas turbines, micro-turbines, photovoltaic (PV) systems, fuel cells, wind-power, combined cooling heat and power (CCHP) systems. DS technologies are batteries, super-capacitors, and flywheels. Renewable energy that comes from resources that are naturally replenished such as sunlight, wind, rain, tides, waves and geothermal heat, are especially promoted in the SG. Solar, wind and hydro power are the three main stream renewable technologies. Solar power is the conversion of sunlight into electricity, either directly using photovoltaics, or indirectly using concentrated solar power (CSP). Air and water flows can be used to rotate turbines that convert kinetic energy into electrical power. Since DER are located close to points of use and often exploit available natural resources or the waste heat from the generation process, they result in many benefits from both economical and operational points of view. Assuming 10 % penetration in the U.S. power grid, distributed generation technologies and smart, interactive storage capacity for residential and small commercial applications could save 10 billion dollars per year by 2020. These technologies will also play significant roles in reducing greenhouse gas emissions and enhancing power quality, reliability and independence [12].

Secondly, in the SG, the efficiency and reliability of power generation and distribution are enhanced by vitalizing the power systems with real-time monitoring and controlling capability. Specifically, computing and digital communications technologies are embedded into grid devices. Intelligent sensors and controllers (e.g., power meters, voltage sensors, fault detectors, circuit breakers, capacitor bank controllers, etc.) are also integrated into the power grid to autonomously collect data and enable remote controls. These intelligent devices are connected to each other and with computing/control centers by data communications networks. As a result, the SG can optimize the operation of its interconnected elements including central and distributed power plants, energy storage stations, transmission and distribution grids, industrial and building automation systems, end-user thermostats, PEVs, appliances and other household devices. Further, inter-networked elements of the SG also enable adaptive system configuration, operation and management that allow the injection of electric energy from alternative sources at any point along the grid. Power energy generated by distributed and small plants using renewable resources can flow into the grid in a well-controlled manner to help relieve the stress placed on the power grid under heavy-load situations. DER, renewable resources along with flexible facilities and mechanisms to incorporate them into the power grid create microgrids that are defined as smart small-scale distributed electricity systems. Microgrids can work either in grid-connected and island modes to achieve specific local goals such as reliability enhancement, carbon emission reduction, diversification of energy sources and cost reduction. They are one of the distinguishing features of the SG.

Thirdly, the SG enhances the efficiency of the power grid by supporting advanced distribution automation (ADA). While traditionally, electric utilities with supervisory control and data acquisition (SCADA) systems have control over the transmission level and distribution level equipment, their area of responsibility does not cover the end-user territory. They are thus unable to provide direct control of smaller energy units such as autonomous DER, homes and buildings. ADA offers extension of utilities' control over these small-scale systems. It incorporates the complete range of functions from existing SCADA systems to smart metering technologies at the customer level, in which local automation, remote control, and central decision making are brought together to deliver a cost-effective, flexible, and cohesive operating architecture [13, 14]. The benefits of ADA can be seen from the following two major perspectives. From the operational and maintenance perspective, ADA can provide improved reliability by reducing outage duration using auto restoration scheme, more efficient voltage control by means of automatic

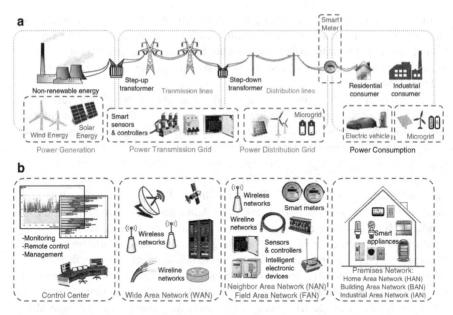

Fig. 1.2 The SG as a revolution of the power grid with new technologies and services. (**a**) Power system. (**b**) Smart grid communications network

Volt-Ampere Reactive (VAR)[1] control, reduced man hour and man power, accurate and useful planning and operational data information, better fault detection and diagnostic analysis, better management of system and component loading. From the customer perspective, it improves service reliability, reduces interruption cost for industrial and commercial customers and provides better power quality. These benefits then in turn result in improved utilization of system capacity, increased revenue due to quick restoration and higher customer loyalty.

Finally, data communications and remote control capability along with the deployment of smart electric meters in the SG enable various intelligent services and applications. Consumers, rather than simply playing a passive role in the existing power grid, can now actively participate in the operations of the grid in order to customize their power consumption profile. A smart meter, one of the key elements that make the power grid smarter, is an electronic device that records consumption of electric energy and communicates that information back to the utility for monitoring, management and billing purposes. It also supports two-way communications between the meter and the central system for other functionalities, which is usually referred to as advanced metering infrastructure

---

[1]VAR is a unit of reactive power of an alternating current, equal to the product of the voltage (in Volts) and the current (in Amperes)

Table 1.1 A comparison of the existing power grid and the SG

Perspective	Today's power grid	The Smart Grid
Generation	- Large-scale plants - Centralized - Far away from consumers - Mainly non-renewable energy sources - Limited connection and cooperation	- Increasing penetration of small-scale plants - Distributed - Closer to consumers - More renewable energy sources for green house effect reduction and energy diversity - Connected and controlled via data communication networks
Transmission	- Tertiary power requested by telephone	- Request of tertiary reserve via broadband connection - Automatic steering of requests with intelligent software - Automatic balancing of power flows
Distribution	- Feed-in follows demand - Limited and local monitoring - Not very efficient control	- Demand automatically adjusts to generation status - Much better knowledge on grid status and power quality - Automatic and optimized control (e.g., fault location, isolation, and service restoration, Volt/VAR optimization)
Metering	- Dump meters - No communications - Meter read by workers	- Smart meters with two-way communications - Two-way communications supporting real-time monitoring and control - Automatic meter reading - Enabling a wide range of smart applications for energy saving and cost reductions
Consumption	- Manual control of loads	- Intelligent monitoring and control of loads - Facilitated with various smart applications for energy saving and cost reduction
Overall operation and management	- Outdated and low-data rate communications equipped to a limited number of elements and subsystems - Limited information - Most functions manually done (measure power quality, search for failures, restore power supply) - Vulnerable to attacks and natural disasters	- Advanced and high-speed communications infrastructure internetworking a vast number of intelligent devices of multiple types from generation to consumption - Continuous information flow from generation to consumption - Monitoring and control in real-time and harmonized manners; many functions are computerized and automatic - Self-healing and resilient to multitudes of attacks and failures

(AMI). With real-time energy consumption information, electricity prices can vary at different times of the day, i.e., time-of-use (TOU) pricing, encouraging consumers to optimize the ways in which they use electricity. Specifically, consumers can adjust the amount or timing of their energy consumption as a response to real-time energy price. For example, the consumers can shift energy consumption from on-peak to off-peak periods. These actions are termed as demand responses in the SG. These responses can effectively help consumers reduce their energy consumption and cost. They can also effectively smooth out energy demand since consumers attempt to lower their consumptions during the peak hours and thus prevent grid overload and cut the costs due to excessive backup capacity in power generation and distribution. A recent Pacific Northwest National Laboratory study in [4] provided homeowners with SG technologies to monitor and adjust the energy consumption in their homes. The average household reduced its annual electric bill by 10 %. If widely deployed, this approach could reduce peak loads up to 15 % annually, which equals to more than 100 GW, or the need to build 100 large coal-fired power plants over the next 20 years in the United States alone. This could save up to 200 billion dollars in capital expenditures on new plant and grid investments, and take the equivalent of 30 million automobiles off the road [4].

The transformation of the existing power grid to the SG is illustrated in Fig. 1.2. Key components, sub-systems and functionalities that shape up the SG are highlighted by dashed blocks. They include smart meters, electric vehicles, microgrids, renewable energy resources, smart sensors and controllers and an underlying communications infrastructure that connects all elements from power generation to consumers and control centers. Table 1.1 summarizes a number of key functions and capability of the SG in contrast to the existing power grid.

1.3 Communications for the Smart Grid

1.3.1 SGCN: Characteristics, Requirements, Challenges

The key to achieve the above-mentioned goals of the SG is the embedding of an advanced communications infrastructure into the power grid. The overall architecture of the SG is therefore usually presented by the integration of the power system and the smart grid communications network (SGCN), as shown in Fig. 1.2. This network transports sensor data and control signals that allow utilities to actively monitor and manage the entire electricity system in a harmonized manner. In other words, the SGCN can be seen as the nervous system of the SG. It is aimed to support all identified SG functionalities including renewable and distributed generation integration, smart metering, demand response, electric vehicle charging, transmission enhancement, advanced distribution automation, consumer-side power management and so on. For example, the SGCN allows utilities to receive information from the grid to determine where power outages or system failures occur as well as their possible causes. It can also send instructions to related devices to prevent cascaded failures or even to fix detected problems. Smart devices in the home, office, or factory inform consumers of the real-time energy price, main power grid status, operations of appliances and allow consumers to dynamically control the injection of power generated by local renewable DER in microgrids for stable operations and cost reductions.

As shown in Table 1.1, most system elements of the SG from power generation to consumption require communications and networking capabilities. It is the SGCN that makes the power grid "smart". Better communications is the key to improve the efficiency and reliability of the power grid, to allow the integration of green and sustainable small-scale energy sources, and to enable value-added smart applications. However, a successful implementation of a cost-effective, efficient, robust and secured SGCN is a challenging task since the SGCN has various characteristics, features, and requirements, different from those of existing residential and commercial communications networks.

Since the SG is a complex system responsible for performing diversified functionalities from power generation, transmission to distribution and consumption, the SGCN interconnects millions of devices of different types in different topologies. For example, one or more smart meters need to be installed for each consumer for information exchange between end-users and utilities. In the U.S. (having more than 115 million households), around 36 million smart meters were already installed by May of 2012 [15]. Due to a huge population of connected devices, the SGCN must be scalable to network size and traffic volume that are expected to increase when more advanced SG applications are emerging in the future. Existing and new devices are upgraded and installed and each of them tend to inject a higher amount of traffic into the network. The SGCN should be able to support network upgrade and expansion with minimal human's intervention. This requires the communications and networking protocols designed for the SGCN to be self-organizing, distributed, scalable, and robust.

Additionally, the SGCN is designed to enable various industrial and residential applications, its constitutive network elements therefore greatly differ in terms of their QoS requirements. For example, home meters typically need to support low-data-rate and non-real-time communications for periodic meter readings and residential demand-response applications. Building or multi-building meters require much higher data rates and near-real-time communications for advanced commercial/industrial applications such as enterprise smart energy management, renewable energy integration, microgrid monitoring and coordination, etc. Devices for critical missions (e.g., fault detection and restoration for transmission and distribution grids) may only need to transmit short messages occasionally, however, these messages require very high reliability and stringent delay. Failures to deliver data or commands necessary for these missions within the allowed time frame compromise the operations of power systems and cause catastrophic consequences. Resource allocation and scheduling mechanisms for efficient provision of a wide range of QoS requirements in the SGCN is therefore challenging. Furthermore, at the present time, the SG is still in its infancy, technical details related to communications traffic profiles and QoS requirements of SG applications have not been well understood. However, due to the fact that most SG applications are for industrial monitoring and control, their characteristics and requirements are quite different from those of today's telecommunications applications such as file transfer, email exchange, web surfing, voice and video streaming.

Even though the SG offers many benefits to the power industry and consumers, its strong dependence on the SGCN obviously makes it vulnerable to cyber threats. Security breaches in the SGCN may result in not only the leakage of consumer information, but also serious consequences ranging from power blackouts, physical damages of equipment and infrastructure, and breakdowns of public safely and national security. As a result, security and privacy issues have been considered by governments, industries, and consumers to be one of the highest priorities for design, deployment, and adoption of the SG. Given the vast scale and complex architecture of the SGCN as well as the presence of various types of traffic and associated QoS requirements, existing security frameworks and protocols employed for public data communications networks are likely no longer adequate. They may need further enhancement or new/additional solutions need to be developed to properly address all security vulnerabilities in the SGCN.

The presence of various communications models is another challenge of the SGCN. Multi-point-to-point (MP2P) is the primary model to be supported. Energy consumption, metering profile, power quality, power line conditions, active and reactive power information, devices status, etc., collected by various types of intelligent devices flow towards data collectors. The converge-cast nature of MP2P traffic requires data processing and communications/networking protocols that are specially developed to efficiently aggregate data and mitigate congestions at collectors. Besides, point-to-multi-point (P2MP) is another important communications to be supported by the SGCN. The information related to real-time pricing, commands for fault isolation or energy flow re-direction, instructions for demand-response functionalities, etc., are conveyed from utility control centers to field devices (smart meters, capacitor banks, voltage regulators, transformers, etc.) for optimizing user energy consumption, smoothing out power consumption peaks, optimizing distribution grid and resolving various failures at distribution level. Point-to-point (P2P) communications between two devices could be exploited to meet security, scalability, and hard real-time requirements in small-scale local segments of the SGCN.

Next, despite the fact that most network elements in the SGCN are static in their locations, the links between them may dynamically change over time. This problem is quite critical considering the fact that many segments of the SGCN are implemented with wireless communications technologies. Multi-path fading can introduce significant variations to wireless link conditions. Surrounding environments (e.g., temporary structures, trees, moving cars or trucks, etc.) and harsh weather conditions (e.g., heavy rain, snow storms, etc.) can also affect the link quality. A typical example is that a big truck can block a network node from a nearby router for a few hours or even longer. Besides, when there is a power outage, a number of nodes have to lower their radio transmission power levels after switching their power supply to battery of limited capacity. As a result, some links may render their connectivity. In order to minimize packet loss and delay, the SGCN needs to have mechanisms to measure or estimate the instantaneous link quality and to be able to adapt well to any link connectivity change.

Most devices of the SGCN are required to operate in outdoor environments without regular maintenance whereas their desired lifetime is 10 years or even 20 years. The communications and networking protocols in this network therefore must be self-healing. Given that the network is composed of millions of nodes, it should be robust to multiple link and/or node failures. For example, a tornado may destroy network elements installed along the power lines, on utility poles, or in substation areas. The SGCN must be able to detect any kind of failures and to have fast actionable responses to them. The traffic needs to be automatically re-routed to alternate paths if primary paths fail or de-toured around regions having problems. Otherwise, the communications will be delayed or corrupted and the SGCN is unable to transport information necessary for detecting and troubleshooting power system faults in the very scenarios where this network is primarily designed to play its roles.

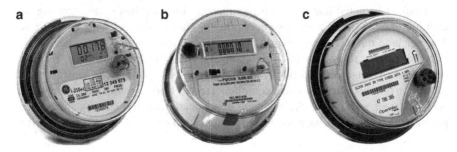

Fig. 1.3 Samples of smart meters manufactured by (**a**) General Electrics [16], (**b**) Landis+Gyr (Toshiba) [17] and (**c**) Itron [18]

1.3.2 Smart Meters

Smart meters (SMs) are one of the most fundamental components of the SG. They are designed to give utilities and end-users more control over electricity distribution, generation, and usage, as well as greater savings and more efficient, reliable services. Unlike conventional electricity meters that can only measure the total amount of electricity used over a billing period, SMs can trace how much and when electricity is used (typically hourly), power quality, power failure, etc., and send this type of information to the utility companies automatically through a communications network. They can also work as a gateway for utility-consumer communications concerning TOU energy pricing, demand response, remote connect/disconnect and load limiting. Especially, in addition to the primary communications interface for information exchange with the utilities, each SM may have another interface to form a local network with smart appliances such as heaters, air conditioners, refrigerators, washers, dryers, PEVs, home energy management system, etc., for demand response related functions.

Each SM typically consists of digital sensors and sub-systems for electrical parameter measurements (e.g., voltage, current, phase, electric power, ...), an embedded computer for data processing and application support, transceivers for information exchange with the utility and appliances, and power-supply units. SMs and metering communications services are manufactured and supplied by numerous electric power and telecom companies including General Electrics [16], Landis+Gyr (Toshiba) [17], Itron [18], S&C Electrics [19], Fujitsu [20], Silver Spring Networks [21], Trilliant [22], Tropos Networks [23], Corinex [24]. Typical examples of SMs are shown in Fig. 1.3. Over the last few years, SMs have been deployed for residential and commercial use by hundreds of utilities in many countries around the world. In the U.S., for example, as of May 2012, 36 million SMs have been installed across the country. As projected in Fig. 1.4, approximately 65 million SMs (more than half of all U.S. households) are expected to be installed by 2015 [15].

It should be noted that the SG is significantly broader than smart metering. In addition to SMs, it incorporates many other connected devices and sub-systems. They include microprocessor-based controllers of power system equipment (e.g., circuit breakers, relays, on load tap changers, capacitor bank switches, recloser controllers, voltage regulators, ...), transmission and power line sensors, tele-protection devices, and so on, that communicate and work together with the electric grid to detect and isolate the outages, dynamically route electricity flows, and perform other functions for power system data acquisition, control and automation.

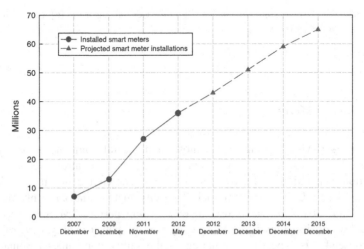

Fig. 1.4 Smart meter installations in the U.S.: 2007–2015 [15] (Source: Institute for Electric Efficiency, Federal Energy Regulatory Commission)

1.4 Structure of This Brief

This Springer Brief is divided into six chapters. This chapter has just covered the basic structure, system elements and key attributes of today's power grid. A number of critical issues and concerns that call for the revolution of the power grid have been addressed. The definition of the SG has been explained. Benefits from economical, operational, management and environmental perspectives of the SG have been presented. A communications infrastructure, so-called the SGCN, as well as its associated requirements that are indispensable for a successful implementation of the SG have been highlighted. More detailed understanding of the SGCN and the investigation of various candidate routing protocols for this network will be presented in subsequent chapters.

Chapter 2 elaborates the SGCN architecture and its representative network segments. Characteristics and required features of each segment are addressed. Next, standards developed for the SG are surveyed. Then, attributes and QoS requirements of different types of traffic anticipated for the SGCN are explored. Chapter 2 provides both the overall structure and technical details of the SGGN that system and network engineers need to grasp when designing and implementing the SGCN.

Chapter 3 surveys a number of representative communications technologies that can be employed for the implementation of the SGCN. It focuses on wireless technologies as they appear to be promising due to their advantages in easy and low-cost deployment, maintenance, and expansion. A typical implementation of the SGSN using multiple communications technologies is illustrated.

Chapter 4 reviews existing routing protocols proposed for wireless mesh networks. Protocols applicable for the neighbor area network (NAN) are focused. Next, main ideas, basic operation principles, weaknesses and strengths of two candidate routing protocols for NANs are presented. Specifically, greedy perimeter stateless routing (GPSR), a representative implementation of location-based routing class, and the routing protocol for low-power and lossy networks (RPL), the state-of-the-art self-organizing coordinate protocol, are considered. Then, a proactive parent switching scheme proposed to deal with network element failures in NANs is explained.

Chapter 5 presents simulation results and related observations that evaluate the feasibility as well as limitations of GPSR and RPL in various practical NAN scenarios. These two protocols are compared in terms of their transmission reliability, latency and adaptivity. The effects of network element failures to the network performance and the benefits of the proposed proactive parent switching scheme are also studies.

Chapter 6 ends this Springer Brief by outlining a number of further technical aspects and issues and their corresponding future work in the SGCN. They include cyber security, service differentiation and provisioning, network coding, machine-to-machine communications, cloud computing, software-defined networking and network virtualization, smart grids and smart cities.

References

1. W. Steinhurst, "The electric industry at a glance," *Silver Spring, MD: National Regulatory Research Institute*, 2008.
2. S. W. Blume, "Electric power system basics: For the nonelectrical professional," *Hoboken, NJ: Wiley-IEEE Press*, 2007.
3. "Electric vehicle geographic forecasts," Navigant Research, Tech. Rep., 2014.
4. C. Feisst, D. Schlesinger, and W. Frye, "Smart grid, the role of electricity infrastructure in reducing greenhouse gas emissions," Cisco internet business solution group, white paper, Oct. 2008.
5. "What caused the power blackout to spread so widely and so fast?" Genscape, Aug. 2003.
6. "Major power outage hits New York, other large cities," CNN, Aug. 2003.
7. "The great 2003 North America blackout," CBC, Aug. 2003.
8. H. L. Willis and W. G. Scott, "Distributed power generation - planning and evaluation," *Marcel Dekker, New York*, 2000.
9. R. G. Pratt *et al.*, "The smart grid: An estimation of the energy and CO2 benefits," U.S. Department of Energy, Tech. Rep., Jan. 2010.
10. Y. Humber, "World needs to get ready for the next nuclear plant accident," Bloomberg, Apr. 2014. [Online]. Available: http://www.bloomberg.com.
11. *Draft Guide for Smart Grid Inter-operability of Energy Technology and Information Technology Operation With the Electric Power System (EPS), and End-Use Applications and Loads*, IEEE P2030 Std., 2011.
12. "Smart grid: Enabler of the new economy," The Electricity Advisory Committee, Tech. Rep., Dec. 2008.
13. E. Clarke, "Control and automation of electrical power distribution systems," *FL:CRC Press*, 2006.
14. D. G. Hart, "How advanced metering can contribute to distribution automation," *IEEE Smart Grid*, Aug. 2012.
15. "Utility-scale smart meter deployments, plans & proposals," Institute for Electric Efficiency, Tech. Rep., May 2012.
16. "General Electrics." [Online]. Available: http://www.gedigitalenergy.com.
17. "Landis+Gyr." [Online]. Available: http://www.landisgyr.com.
18. "Itron." [Online]. Available: http://www.itron.com.
19. "S&C Electrics." [Online]. Available: http://www.sandc.com.
20. "Fujitsu." [Online]. Available: http://www.fujitsu.com.
21. "Silver Spring Networks." [Online]. Available: http://www.silverspringnet.com.
22. "Trilliant." [Online]. Available: http://www.trilliantinc.com.
23. "Tropos Networks." [Online]. Available: http://www.tropos.com.
24. "Corinex." [Online]. Available: http://www.corinex.com.

Chapter 2
Smart Grid Communications Network (SGCN)

This chapter provides a top-level system description as well as technical details for the SGCN so as to help readers grasp system and network requirements and challenges when designing and implementing the SGCN. Specifically, this chapter elaborates on the overall architecture of the SGCN by decomposing it into three representative network segments. For each segment, details regarding required communication delay, bandwidth, network coverage and potential applications are addressed. Since inter-operability is one of the most vital concerns in the SGSN, this chapter then gives an overview on the standards for the SGCN developed by various organizations such as the Institute of Electrical and Electronics Engineers (IEEE) and the National Institute and Technology (NIST). Finally, this chapter discusses QoS attributes and requirements of the various elements in the SGCN.

2.1 Overall Architecture of the SGCN

The SGCN is typically composed of various segments, each of which is responsible for information and control message exchanges within a specific region of the power grid as sketched in Fig. 1.2b. Communications characteristics of these segments will be discussed in the following subsections.

2.1.1 Premises Network

The premises network gathers sensor information from a variety of smart appliances and devices within the customer premises and delivers control information to them for better energy consumption management. The coverage areas of this network could be apartments, homes, residential/commercial buildings, and factories.

© The Author(s) 2014
Q.-D. Ho et al., *Wireless Communications Networks for the Smart Grid*,
SpringerBriefs in Computer Science, DOI 10.1007/978-3-319-10347-1_2

2.1.1.1 Home Area Network (HAN)

A HAN is deployed in an apartment or a residential dwelling. It can support functions such as cycling heaters, washers/dryers, or turning air conditioners off during peak load conditions and controlling the charging/discharging procedure for PEVs. An important component of a HAN is the home energy management system (HEMS) that allows consumers to see how much power their household is consuming at any moment in time as well as over a period of time. In order to facilitate applications related to TOU-based energy management, demand response, etc., the HEMS communicates with a smart meter (SM) installed in consumer site and works as a communications gateway relaying information related to real-time energy price, home energy usage information, and control signals between the HAN and the utility. HEMS allows the consumer to customize their power usage profile in order to minimize their electricity bill. Typically, HANs need to cover areas of up to $200\,m^2$ and support from 10 to 100 kilobits per second (kbps).

2.1.1.2 Building Area Network (BAN)

Similar to HAN, a BAN is responsible for monitoring and controlling consumer smart devices and exchanging information with the utilities. However, it needs to cover an entire building which consists of multiple apartments and offices. A BAN can be a collection of HANs connected with a building SM which is typically installed at the building's power feeder. Especially, the BAN may incorporate a microgrid that generates electricity by harvesting heat wastes or renewable resources such as solar or wind energy. Due to a higher number of network elements and energy management applications, it requires higher data rates, compared to HAN.

2.1.1.3 Industrial Area Network (IAN)

An IAN is a communications network deployed in factory floors. It incorporates connected sensors, controllers, and specialized building management software. The IAN handles building or multi-building applications, such as building automation systems and energy management, for optimizing energy, economic and environmental performance of all connected devices. Similar to BAN, a microgrid is also an important element of IAN. However, a microgrid of this network has a larger scale, higher capacity and complexity than that of BAN. Additionally, as industrial customers run more sophisticated applications, their SMs should possess the ability to record additional data such as power quality, voltage sags/surges, and phasor measurements.

Despite the fact that there are a number of differences between HAN, BAN, and IAN as just mentioned, these networks share many common characteristics and design disciplines. They are mostly deployed in indoor environments and need to support short-range communications between network elements for monitoring and control applications. Also, they use a SM as a gateway to connect them with other network segments and the utilities. As a result, thereafter, the HAN is used to represent the premises network.

2.1.2 Neighbor Area Network (NAN)

The NAN is responsible for smart metering communications that enables information exchange between customer premises and utility company's WANs. NAN endpoints are SMs that are considered to be at the heart of SG revolution. SMs support energy consumption recording and real-time or near real-time data acquisition and control for various SG applications including distribution automation, power outage management, power quality monitoring, etc. A NAN cluster usually covers an area of several square kilometers. The number of SMs in each cluster varies from a few hundreds to a few thousands depending on the power grid topology and the employed communications technology and protocol. The data rate required by each SM may widely vary depending on deployed applications. For example, for interval and on-demand meter reading, only around a few bps per meter is required. However, in order to support future applications, such as advanced distribution automation, fault detection and restoration and so on, higher data rates, e.g., a few tens of kbps per meter, may be required. It is noted that the NAN is a critical segment of the SGCN since it is responsible for transporting a huge volume of different types of data and distributing control signals between utility companies and a large number of devices installed at customer premises.

2.1.3 Field Area Network (FAN)

A FAN provides connectivity for smart devices in transmission and distribution grids and substations. These devices include power line monitors, breaker controllers, voltage regulators, capacitor bank controllers, recloser controllers, smart transformers, data collectors, etc. They are used to quickly detect anomalies and failures and to automate responses to improve reliability and quality of power services. Besides, the FAN enables mobile workers to access field devices using their laptops, tablets or hand-held equipment in order to collect and analyze data for failure/fault detection, troubleshouting and service restoration. Similar to NAN, this network segment incorporates a large number of devices and covers wide areas. NAN and FAN may also have overlapped coverage since numerous smart devices are tied to both of them for successful implementations of various emerging applications. As an illustrative example, SMs need to be accessible by both of these network segments to ensure that the distribution grid can obtain vital information from customer premises in real-time to enable efficient vol/VAR control.

Therefore, NAN and FAN share many design principles and communications technologies. It is sufficient to only focus on the NAN that can be considered as a representation of these two segments.

2.1.4 Wide Area Network (WAN)

A WAN aggregates data from multiple NANs and conveys it to utility company's private networks. It also enables long-haul communications among different data aggregation points (DAPs) of power generation plants, distributed energy resource stations, substations, transmission and distribution grids, control centers, etc. Additionally, the utility company's WAN is responsible for providing the two-way network, needed for substation communications, distribution automation, power quality monitoring, etc., while also supporting data aggregation and back-haul for NANs. The WAN may cover a very large area, i.e., thousands of square kilometers and could aggregate a large number of supported devices and thus require hundreds of megabits per second (Mbps) of data transmission.

2.1.5 Interconnection of Network Segments

In order to form the SGCN, the network segments that have been just presented in the above subsections are interconnected through gateways: a SM between HAN and NAN and a DAP between NAN and WAN. A SM collects the power-usage data of a home or building by communicating with the home network gateway or functioning as the gateway itself. The DAP aggregates data from a cluster of SMs and relays it to the grid operator's control centers. Instructions for optimizing the power grid and user energy consumption can be sent from control centers to intelligent electronic devices (IEDs) and consumer devices through WAN, NAN and HAN in the opposite direction. These segments may employ different communications technologies and protocols to meet their own requirements in terms of data rates, communications latencies, deployment/maintenance costs. Therefore, in addition to data aggregation/filtering and traffic routing, the gateways also perform network address translation, protocol translation/mapping, etc., as necessary to provide system inter-operability.

2.2 Standards in the SGCN

It has been observed that inter-operability is at the heart of technological revolutions in telecommunications, transportation, industrial manufacturing, and many other industries. As presented in the preceding section, the SGCN is the integration of multiple segments, each of which is required to provide network connectivity for a vast number of devices of different types. Therefore, inter-operability in this network is also a vital concern.

By definition, inter-operability is "the ability of two or more systems or components to exchange information and to use the information that has been exchanged" [1]. The lack of widely accepted standards limits the inter-operability between SMs, smart monitoring and controlling devices, renewable energy sources and emerging advanced applications, and thus prevents their integration. As a result, many regional, national, and international Standards Development Organizations (SDOs) have been working towards a variety of standards for the SG, e.g., Institute of Electrical and Electronics Engineers (IEEE) [2], National Institute of Standards and Technology (NIST) [3], American National Standards Institute (ANSI) [4], International Electrotechnical Commission (IEC) [4], International Organization for Standardization (ISO) [5], International Telecommunication Union (ITU) [6], etc. In addition to SDOs, there are a number of alliances that recognize the value of a particular technology and attempt to promote specifications as standards for that technology. For example, some well-known alliances related to the utility industry in the HAN market are ZigBee Alliance [7], WiFi Alliance [8], HomePlug Powerline Alliance [9], Z-Wave Alliance [10]. The following subsections provide an overview of the key roles and activities of the IEEE and the NIST in developing standards for inter-operability of the SG.

2.2.1 IEEE Standards

IEEE has more than 100 approved standards and many under-development standards relevant to the SG. This organization is also working closely with the NIST and other standards bodies in developing a standard roadmap and conformance testing and certification framework for the SG [11]. The IEEE P2030 [12] is a standard guide for SG inter-operability. It provides understanding, definitions, and guidance for design and implementation of SG components and end-user applications for both legacy and future infrastructures. The knowledge base addresses terminology, characteristics, functional performance and evaluation criteria, and the application of engineering principles for SG inter-operability of electric power system (EPS) with end-use applications and loads. Besides, the reference model, namely Smart Grid Inter-operability Reference Model (SGIRM), presents three different architectural perspectives with inter-operability tables and charts. The IEEE 2030 series of standards will address more specific technologies and implementation of SG systems (e.g., P2030.1 Electric Vehicle, P2030.2 Storage Energy Systems).

The SGIRM is the central part of the IEEE P2030 standard. It is intended to present inter-operable design and implementation alternatives for systems that facilitate data exchange between SG elements, loads, and end-use applications. The IEEE P2030 SGIRM encompasses conceptual architectures of SG from power systems, communications, and information technology perspectives and characteristics of the data that flows between the entities within these perspectives. Each conceptual architecture presents a set of labeled diagrams that offer standards-based architectural direction for the integration of energy systems with information and communications technology (ICT) infrastructures of the evolving SG. It aims to establish a common language and classification for SG community to communicate effectively. The interfaces between entities in each architecture will typically contain a wide variety of data. The IEEE SGIRM data classification reference table provides guidance in identifying a set of characteristics for the data at those interfaces. It is a starting point in determining appropriate classifications for the data.

The Inter-operability Architectural Perspectives (IAPs) primarily relate to logical, functional considerations of power systems, communications, and information technology interfaces for SG inter-operability. The Power Systems IAP (PS-IAP) mostly represents a traditional view of the EPS, while Communications Technology IAP (CT-IAP) provides a means to getting the data from place to place and the Information Technology IAP (IT-IAP) provides a means to manipulate data to provide useful information. A summary of the three perspectives is as follows:

- *PS-IAP*: The emphasis of the power system perspective is the production, delivery, and consumption of electric energy including apparatus, applications, and operational concepts. This perspective defines seven domains common to all three perspectives: bulk generation, transmission, distribution, service providers, markets, control/operations, and customers.
- *CT-IAP*: The emphasis of the communications technology perspective is communication connectivity among systems, devices, and applications in the context of SG. The perspective includes communications networks, media, performance, and protocols.
- *IT-IAP*: The emphasis of the information technology perspective is the control of processes and data management flow. The perspective includes technologies that store, process, manage, and control the secure information data flow.

The IEEE SGIRM data classification reference table presents various data characteristics (e.g., reach, information transfer time, latency, etc.) and their corresponding value ranges (representative of values that are typically used). The user of the table may need to identify more appropriate data characteristics and values for their specific circumstances.

Besides, the IEEE P2030 SGIRM methodology provides understanding, definitions, and guidance for design and implementation of SG components and end-use applications for both legacy and future infrastructures. The key to using the IEEE P2030 SGIRM is to determine the relevant interfaces, data flows, and data characteristics based on the intended SG application requirements and goals. Once the data requirements of the goals have been defined, the users, based on SGIRM, select a set of interfaces on each IAP of the model that meet the data needs. These interfaces and data flow characteristics are key elements for subsequent SG architectural design and design of implementation operations. The determination of these interfaces is the first step toward determining the implementation of the intended SG application requirements and goals. To assist in this step, the PS-IAP interface tables are provided to identify logical information to be conveyed, the CT-IAP interface tables identify the general communication options of the interface, and the IT-IAP data flow tables identify the general data types.

2.2.2   NIST Standards

NIST has been assigned the "primary responsibility to coordinate development of a framework that includes protocols and model standards for information management to achieve inter-operability of smart grid devices and systems ..." (Energy Independence and Security Act of 2007, Title XIII, Section 1305). Its primary responsibilities include (i) identifying existing applicable standards, (ii) addressing and solving gaps where a standard extension or new standard is needed and (iii) identifying overlaps where multiple standards address some common information. NIST has developed a three-phase plan to accelerate the identification of an initial set of standards and to establish a robust framework for the sustaining development of the many additional standards that will be needed and for setting up a conformity testing and certification infrastructure.

NIST Framework and Roadmap for Smart Grid Inter-operability Standards, Release 1.0 [13], is the output of the Phase I of NIST plan. It describes a high-level conceptual reference model for SG, identified 25 relevant standards (and additional 50 standards for further review) that are applicable to the ongoing development of SG and described the strategy to establish requirements and standards to help ensure SG cybersecurity. Release 2.0 of NIST Framework and Roadmap for Smart Grid Inter-operability Standards [14] details progress made in Phases II and III. Major deliverables have been produced in the areas of SG architecture, cybersecurity, and testing and certification. Release 2.0 [14] presented 34 reviewed standards (and additional 62 standards for further review). The listed standards have been undergone an extensive vetting process and are expected to stand the "test of time" as useful building blocks for firms producing devices and software for SG. Ongoing standards coordination and harmonization process carried out by NIST will ultimately deliver communications protocols, standard interfaces, and other widely accepted and adopted technical specifications necessary to build an advanced, secure electric power grid with two-way communications and control capabilities [14]. Release 3.0 of NIST Framework and Roadmap for Smart Grid Inter-operability Standards [15] updates NIST's ongoing efforts to facilitate and coordinate smart grid inter-operability standards development and smart grid-related measurement science and technology, including the evolving and continuing NIST relationship with the Smart Grid Inter-operability Panel (SGIP). Lists of standards approved and under-reviewed by NIST can be found in [13–15]. For examples, IEC 61850 protocol suite is for communications within transmission and distribution sectors; ANSI C12.20 is for revenue metering accuracy specification; and IEEE 1588 is for time management and clock synchronization of equipments across the SG.

Table 2.1 PAPs identified by NIST

Supporting	PAPs
Metering	Meter upgradeability standard (PAP 00)
	Standard meter data profiles (PAP 05)
	Translate ANSI C12.19 to the common semantic model of common information model (CIM) (PAP 06)
Enhanced customer interactions with the SG	Standards for energy usage information (PAP 10)
	Standard demand response signals (PAP 09)
	Develop common specification for price and product definition (PAP 03)
	Develop common scheduling communication for energy transactions (PAP 04)
Smart grid communications	Guidelines for the use of IP protocol suite in SG (PAP 01)
	Guidelines for the use of wireless communications (PAP 02)
	Harmonize power line carrier standards for appliance communications in the home (PAP 15)
Distribution and transmission	Develop CIM for distribution grid management (PAP 08)
	Transmission and distribution power systems model mapping (PAP 14)
	IEC 61850 objects/distributed network protocol 3 (DNP3) mapping (PAP 12)
	Harmonization of IEEE C37.118 with IEC 61850 and precision time synchronization (PAP13)
New smart grid technologies	Energy storage interconnection guidelines (PAP 07)
	Inter-operability standards to support plug-in electric vehicles (PAP 11)

In addition to reviewing and selecting applicable standards for SG, NIST has another important contribution in identifying a set of Priority Action Plans (PAPs) for developing and improving standards necessary to build an inter-operable SG. Those PAPs arise from the analysis of the applicability of standards to SG use cases and are targeted to resolve specific critical issues. Each PAP addresses one of the following situations: a gap exists, where a standard extension or new standard is needed; an overlap exists, where two complementary standards address some information that is in common but different for the same scope of application [13–15]. A number of representing PAPs is summarized in Table 2.1.

As an illustrative example, PAP 02 deals with wireless communications for SG. It provides key tools and method to assist SG system designers in making informed decisions about wireless technologies. An initial set of quantified requirements has been brought together for AMI and initial DA communications. This work area investigates the strengths, weaknesses, capabilities, and constraints of existing and emerging standards-based physical media for wireless communications. The approach is to work with the appropriate SDOs to determine the characteristics of each technology for SG application areas and types. Results are used to assess the appropriateness of wireless communications technologies for meeting SG applications. A complete list of PAPs addressed by NIST can be found in [13–15].

2.3   QoS Requirements in the SGCN

The SGCN is designed for large-scale emerging SG industrial applications. There-fore, its anticipated traffic is likely to be quite different from that generated by commercial and enterprise communications networks in use today. Specifically, the SGCN has to be robust and secure. High network availability is critical along with predictable sub-second convergence for any failures. The network should possess a degree of fault tolerance for increased resiliency and have the ability to self-recover. Additionally, the network should support a secure end-to-end transport layer ensuring confidentiality, integrity and privacy of the data for meeting North American Electric Reliability Corporation-Critical Infrastructure Protection (NERC-CIP) regulatory requirements [16, 17].

However, specific requirements vary based on the nature and objectives of the deployed SG application. For example, critical information required for stable and reliable grid operation will be time sensitive and thus have stringent latency requirements. Furthermore, even specific SG applications may require multiple priority settings based on the context of the grid operation. For instance, the desired QoS differentiation for periodic meter reads will vary based on whether the grid is operating in a conventional manner, during an outage or with other active applications that need real-time information (e.g., demand response). Therefore, the SGCN is faced with two important QoS factors: a wide range of latency, bandwidth, security and reliability requirements and the need for dynamic flow priority associations based on grid condition and operation [19].

With that in mind, an in-depth description of anticipated SG applications, extracted from various technical documents including [16–23], is presented below where the applications are classified based on their network association. Specifi-cally, the three main groups are (i) home and AMI, (ii) substation networks and (iii) distribution networks. Further, Table 2.2 gives a summary of the various types of SGCN traffic and their respective bandwidth and latency requirements.

2.3.1   Home and AMI Networks

Home and AMI network applications handle the two-way communication between the consumer and the SG. In the uplink direction, from consumer to control center, application communications can range from periodic meter reads to failure notifications. In the downlink direction, from control center to consumer, appli-cations can allow for optimization of electricity usage. In particular, three main classifications can be presented: (i) electricity usage applications, (ii) electric grid state applications, and (iii) demand optimization applications.

Table 2.2 SGCN traffic types and their required QoSs

Traffic type	Traffic regularity and data rate	Latency
Home and AMI networks		
In-home communications	Regular/on-demand A few kbps per device	$2 \sim 15$ s
Meter reads	regular/on-demand A few bps $\sim$ kbps per meter	$2 \sim 15$ s; 100's of ms (for advanced applications)
Connects and disconnections	Occasional Very low rate	Long (customer moving); 100's ms (fast responses to grid conditions)
Outage management	Occasional Low rate	Near real-time (10's of ms)
Demand response (DR)	Occasional/on-demand 10's of kbps	500 ms (mission-critical) up to several minutes (load balancing)
Power trading information	Periodical Low rate	10's of seconds
Substation networks		
Synchrophasor	Occasional/on-demand $600 \sim 1,500$ kbps	$20 \sim 200$ ms (monitoring and control); Long (historical data)
SCADA	Polling $10 \sim 30$ kbps	$2 \sim 4$ s
Inter-substation	Regular Variable rate	$12 \sim 20$ ms
Site surveillance	Periodical/event-triggered A few Mbps	A few seconds
Distribution network		
FLIR	Event-triggered $10 \sim 30$ kbps	Real-time
Distribution automation	Periodical A few Mbps	$25 \sim 100$ ms
Event notification signals	Occasional/event-triggered Burst of data	Near real-time
Asset management	Periodical/on-demand Variable rates	Variable latencies
Workforce access	Occasional 250 kbps or higher	150 ms or lower

2.3.1.1 Electricity Usage Applications

At the home level, usage monitoring applications can be used to transfer instantaneous electricity usage for each device to the SM. The transmitted data is typically only a few kbps per device and the latency is not critical and could be between 2 and 15 s [18, 22, 23]. With this information, HAN applications can then optimize home electricity usage and thereby reduce overall home power consumption.

At the neighbor level, aggregate energy consumption information is transmitted by SMs (for each home) on a periodical basis. The associated traffic is predictable and has long latency requirements. For conventional meter readings, only basic power use information is considered and thus the required data rate is very low, i.e., only a few bits per second (bps) per meter, and the latency is in the range of 2–15 s. As they are done on a periodic basis, only medium reliability is required but high security is still necessary to ensure a safe and secure SG [19]. However, for advanced applications (e.g., power quality monitoring, advanced distribution automation, etc.), many other parameters (e.g., active and reactive power, phase and frequency) need to be collected at much higher frequencies. Each meter may therefore need higher data transmission rate and require more stringent latency [16, 18]. For instance, for critical and priority AMI data, based on grid operating conditions, the delay allowance drops to 250 and 300 ms, respectively [21–23].

2.3.1.2   Electric Grid State Applications

The state of the electric grid can change with either scheduled modifications in consumer connects/disconnects or with failures. In the case where customers move, the change to the grid state is in response to a planned event and thus long latencies are tolerated and the triggered connect/disconnect does not affect overall grid performance. However, when connect/disconnect operations are used as responses to grid conditions, in order to ensure grid stability and reliability, the required latency may drop to only a few hundred milliseconds [24].

Further, in the event of a power outage, i.e., short circuits, failures at power stations and damage in transmission and/or distribution lines [25], fast response is necessary. Traditionally, outages are reported via phone calls from customers. However, to enable fast outage detection and recovery, the SG allows for outage management systems (OMSs) that are used to predict outage location, provide outage analysis and allow for service restoration. Moreover, the OMS can be enhanced with the inclusion of near real-time data exchanged between SMs and control centers. With this integration, SMs can act as a trip wire to indicate the loss of power at an end-point. They can be programmed to automatically give a "last gasp" message to indicate that they have lost power, thereby providing the utility company with valuable information for pin-pointing the origins of the outage [26]. Additionally, with this notification system, utilities can forego the extra manpower required for accurate outage reports and analysis. As outage management is a critical function of the SG, this type of message falls under the critical AMI setting and requires latency within the 250-ms range to ensure grid reliability.

2.3.1.3   Demand Optimization Applications

As the SG allows for near real-time electricity consumption information, it includes the ability to optimize electricity usage. For example, the Demand Response (DR) application allows utilities to communicate with home devices such as load controllers, smart thermostats and home energy consoles in an attempt to reduce or shift power use during peak demand periods and thereby mitigate the need for rolling blackouts. In particular, with direct load control, this power usage shift can be triggered by a simple switch-off command to an appliance and thus its bandwidth requirement is quite low, i.e., few tens of kbps [18]. Estimates of the latency requirements of DR fall into a wide range, from as little as 500 ms (e.g., for mission-critical control messages) up to several minutes (e.g., for load balancing management) [16, 18].

Furthermore, customers can participate through demand pricing. Specifically, the nodal market price for power will vary every 5 min and customers opting for dynamic power pricing can buy their power under current market conditions. This means that a water heater, for example, would receive the information and could use it to decide when to run and when to remain idle. All nodal pricing will need to be available in a centralized manner in one place for some market traders. Others will just want selective data. The exact format of this information is unknown at this time, but it is expected that individual nodal price updates will be small, perhaps 1,400 bytes in size.

2.3.2   Substation Networks

At the substation level, substation automation systems (SAS) are designed for monitoring, control and protection of substation devices. These applications perform actions based on collected real time data. To that effect, communications in this setting is critical and should be highly reliable, scalable, secure and cost-effective [25]. As for communications between substations, emerging applications such as distributed energy resources and distribution automation rely on communications with strict latency requirements from 12 to 20 ms [16, 22, 23]. Specifically, most substation SG applications can be categorized as either monitoring or control applications, some examples are presented below.

2.3.2.1   Monitoring Applications

Wide area situational awareness (WASA) refers to the implementation of a set of technologies designed to improve the monitoring of the power system across large geographic areas and thus respond to power system disturbances and cascading blackouts in an efficient manner. One of its primary measurement technologies is synchrophasor.

Synchrophasor traffic has varying levels of latency requirements. For real-time monitoring and control, latency requirements are very stringent, i.e., from 20 to 200 ms. Specifically, latency requirements are in the range of 60 ms for measurements, 100 ms for phasor measurement units (PMUs) clock synchronization and 500 ms for PMU data. For post-event, historical data, low latency is less imperative [18, 22, 23, 27]. The required bandwidth is between 600 and 1,500 kbps and its main factors are the number of PMUs, word length, number of samples and frequency [16, 18, 22, 23]. Additionally, synchrophasors require a more stringent reliability of approximately 99.99995 % which equates to being out of service for $16\,s\,year^{-1}$ [18].

Additionally, monitoring of transmission lines is crucial for detecting icing, overheating and lightning strikes. In this case, the monitoring scheme includes deploying wireless sensor nodes on some transmission line parts and using relays to gather the transmission line condition information. However, the specifics of communications requirements vary based on the network model, the number of nodes and the preferred communication technology [25].

Furthermore, substation surveillance applications are proposed for enhanced security. These applications require high bandwidths of up to a few Mbps, especially for video surveillance, and the primary factors for bandwidth usage are the number of cameras and the video's resolution. This traffic type can tolerate latencies of a few seconds [16].

2.3.2.2 Control Applications

The standard substation control application is SCADA or Substation Supervisory Control and Data Acquisition. It considers the traffic generated when the master periodically polls IEDs inside the substation. The required bandwidth depends on the number of polled devices and it is forecasted to be around 10–30 kbps. The latency requirement is typically from 2 to 4 s [16, 17]. However, under certain grid conditions, latency requirements are more stringent. For instance, load shedding for underfrequency has a delay allowance of only 10 ms, SCADA critical measurements for poll response require 100 ms, most distribution and SCADA applications require 250 ms. In the second range, SCADA applications include image files, fault recorders, medium speed monitoring and control information, low speed observation and measurement information, text strings, audio and video data streams [21]. Additionally, high security and reliability are required [19].

2.3.3 Distribution Network

At the distribution level, SG applications are mainly employed for two main objectives. First, distribution network communications can be used to detect failure events. Second, communications in this segment can be incorporated to optimize electricity distribution, utility assets and even workforce access.

2.3.3.1 Grid State Applications

When communications are incorporated to detect failures, data will be sent from event/fault recorders whenever an event occurs. An event, for example, might be a lightning strike followed by a set of circuit breakers that trip in response. Sampled waveforms of a number of voltages and currents at 5 kHz for seconds are possible. This data can be quite large when compared to many of the other types of information that are passed around the system. These files will be sent after the fault has occurred, meaning that they normally do not interfere with the current situation. However, if a fault occurs, and it is followed by another fault, then interference could occur. Similarly, if a line is faulted and a device, known as an auto recloser, attempts to reconnect the line, then this could cause a second fault.

When these notifications are sent, the Fault Location, Isolation and Restoration (FLIR) application is used to restore the grid. Since this application is related to grid stability, it has very low latencies requirements in the range of a few milliseconds. Specifically, high speed protection information requires 8–10 ms of delay. Breaker reclosures, lockout functions and many transformer protection and control applications need 16 ms. Finally, some lower priority protection and control applications can tolerate latencies of up to 500 ms and 1 s [21]. The primary factors for its bandwidth usage include the circuit complexity and number of communication steps involved before the fault can be isolated. FLIR typically requires from 10 to 30 kbps [28].

2.3.3.2 Distribution Optimization

Distribution automation (DA) is the service that deals with volt/var and power quality optimization on the distribution grid. In particular, it optimizes the flow of electricity from the utilities to consumers in order to enhance the efficiency and reliability of power delivery. Generally, wide-spread inclusion of DA is expensive but it becomes more important in scenarios with distributed energy resources. This service may generate from 2 to 5 Mbps of traffic and require 25–100 ms of delay bound [29]. Further, high security and reliability are necessary [19, 22, 23].

Asset management is the service for predictively and pro-actively gathering and analyzing non-operational data for potential asset failures. Specifically, it offers management, automation, tracking, and optimization of the work order process, field crew scheduling and field assets [25]. Further, with the introduction of "smart" sensors and monitoring equipment that allow for communications, asset management systems can balance the performance of the system, avert risk of failure and enhance reliability. With that in mind, the primary drivers for bandwidth in this case are the number of assets and the amount of non-operational data that needs to be monitored to predict the health of the asset. As for workforce access, it provides expert video, access to local devices and voice communications with field workers. It typically requires 250 kbps of bandwidth and 150 ms of latency [16, 17, 22, 23]. Specifically, when considering the mobile workforce, latency requirement for enterprise data is around 250 ms, while those for real-time video and push-to-talk Voice over Internet Protocol (VoIP) bearers/signaling are around 200 and 175–200 ms, respectively [21].

As for workforce access, it provides expert video, access to local devices and voice communications with field workers. It typically requires 250 kbps of bandwidth and 150 ms of latency [16, 17, 22, 23]. Specifically, when considering the mobile workforce, latency requirement for enterprise data is around 250 ms, while those for real-time video and push-to-talk VoIP bearers/signaling are around 200 and 175–200 ms, respectively [21].

References

1. *IEEE Standard computer dictionary: A compilation of IEEE standard computer glossaries,* Institute of Electrical and Electronics Engineers Std., 2003.
2. "Institute of Electrical and Electronics Engineers." [Online]. Available: http://www.ieee.org.
3. "National Institute of Standards and Technology." [Online]. Available: http://www.nist.gov.
4. "American National Standards Institute." [Online]. Available: http://www.ansi.org.
5. "International Organization for Standardization." [Online]. Available: http://www.iso.org.
6. "International Telecommunication Union." [Online]. Available: http://www.itu.int.
7. "ZigBee Alliance." [Online]. Available: https://www.zigbee.org.
8. "WiFi Alliance." [Online]. Available: http://www.wi-fi.org.
9. "Home Plug Poweline Alliance." [Online]. Available: https://www.homeplug.org.
10. "Z-Wave Alliance." [Online]. Available: http://www.z-wavealliance.org.
11. "IEEE Smart Grid." [Online]. Available: http://smartgrid.ieee.org/standards.
12. *Draft Guide for Smart Grid Inter-operability of Energy Technology and Information Technology Operation With the Electric Power System (EPS), and End-Use Applications and Loads,* IEEE P2030 Std., 2011.
13. *NIST framework and roadmap for smart grid inter-operability standards, release 1.0,* Special Publication 1108, National Institude of Standards and Technology Std., 2010.
14. *NIST framework and roadmap for smart grid inter-operability standards, release 1.0,* Special Publication 1108R2, National Institude of Standards and Technology Std., 2012.
15. *Preleminary discussion draft - NIST framework and roadmap for smart grid inter-operability standards, release 3.0,* Special Publication 1108R3, National Institude of Standards and Technology Std., 2014.
16. Y. Gobena et al., "Practical architecture considerations for smart grid WAN network," in *Proc. Power Systems Conference and Exposition (IEEE/PES),* 2011, pp. 1–6.
17. K. Hopkinson et al., "Quality-of-service considerations in utility communication networks," *IEEE Transactions on Power Delivery,* vol. 24, no. 3, pp. 1465–1474, Jul. 2009.
18. "Communications requirements of smart grid technologies," US's Department of Energy, Tech. Rep., Oct. 2010.
19. E. W. Gunther et al., "Smart grid standards assessment and recommendations for adoption and development," EnerNex Corporation, Tech. Rep., 2009.
20. *IEEE standard communication delivery time performance requirements for electric power substation automation,* IEEE 1646-2004 Std., 2011.
21. G. D. Jayant et al., "Differentiated services QoS in smart grid communication networks," Alcatel-Lucent, Tech. Rep., 2011.
22. Q. D. Ho, Y. Gao, and T. Le-Ngoc, "Challenges and research opportunities in wireless communications networks for smart grid," *IEEE Wireless Communications,* pp. 89–95, Jun. 2013.

23. Q. D. Ho and T. Le-Ngoc. *Smart grid communications networks: Wireless technologies, protocols, issues and standards*, Chapter 5 in Handbook on Green Information and Communication Systems (Editors: S. O. Mohammad, A. Alagan, and W. Isaac), Elsevier, Summer 2012.
24. "Distribution operations curtails customer load for grid management," SCE Internal, Jan. 2010.
25. V. C. Gungor *et al.*, "A survey on smart grid potential applications and communication requirements," *IEEE Trans. on Industrial Informatics*, vol. 9, no. 1, pp. 28–42, Feb. 2013.
26. "Outage management: The electric utilityŠs no.1 headache," TROPOS networks, Tech. Rep., Jul. 2007.
27. "Sychrophasor architectural issues," NASPINET Initiative.
28. G. Hataway, T. Warren, and C. Stephens, "Implementation of a high-speed distribution network reconfiguration scheme," in *Proc. Power Systems Conference: Advanced Metering, Protection, Control, Communication, and Distributed Resources*, Mar. 2006, pp. 440–446.
29. "Voltage regulation and protection issues in distribution feeders with distributed generation," SCE, Jan. 2010.

Chapter 3
Wireless Communications Technologies for the SGCN

As the enabling technology of the SG, a viable design for the SGCN is crucial. Further, the choice of communications technology per segment will greatly alter deployment and maintenance costs as well as overall performance. With that in mind, in this chapter, an in-depth survey of representative communications technologies with a focus on wireless technologies is presented. Then, a typical implementation of the SGCN is shown with IEEE 802.15.4, IEEE 802.11 and cellular networks for the HAN, NAN and WAN, respectively.

3.1 Brief Overview of Communications Technologies for the SGCN

Numerous wireline and wireless communications technologies that have been developed for residential, industrial and enterprise applications can be used to implement the SGCN. For wireline technologies, candidates are digital subscriber line (DSL), leased line, power line communications (PLC), fiber optics and so on. DSL is a family of technologies that provide data access over a local telephone network. The data rate of consumer DSL services typically ranges from 256 kbps to 40 Mbps depending on DSL technology, line conditions, and service-level implementation. Leased line is a private circuit or data line permanently connecting two or more locations. For example, a leased-line T1 provides a dedicated maximum transmission rate of 1.544 Mbps. As opposed to DSL, a leased line is always active and not shared with other users. Therefore, it can assure a given level of quality at the cost of higher service fees. PLC carries data over power cables originally used for electric power transmission and/or distribution. Data rates supported by PLC vary from a few kbps to Mbps depending on technologies, cable conditions and communications distances. Optical fiber communications carry data reliably at a very high data rate, in the range of Gbps, using light signals. One of the main

© The Author(s) 2014

Q.-D. Ho et al., *Wireless Communications Networks for the Smart Grid*,
SpringerBriefs in Computer Science, DOI 10.1007/978-3-319-10347-1_3

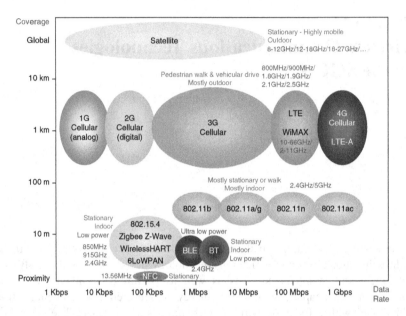

Fig. 3.1 An overview of wireless communications technologies

disadvantages of wireline communications technologies is their high installation and maintenance costs. Also, it is quite challenging to run new cables in some situations, e.g., in an existing buildings or to reach distant devices. DSL and leased line require monthly service fees. PLC appears cost-effective; however it might not have widespread use due to several shortcomings including the existence of various sources of noise/interference in power lines, significant signal attenuation/distortion when crossing transformers, limited data rates, etc. [1, 2]. Fiber optic communications is expensive. Therefore, wireline technologies are economically feasible for the SGCN when network cables and related facilities are pre-existing and readily available at acceptable costs. They tend to be more suitable for back-haul links carrying a large volume of traffic between data aggregation points (DAPs) and control centers. For example, optical fiber is usually available in most substations and can be used to connect DAPs at substations and utilities. New residential and commercial buildings often have gigabit fiber links that can be used to connect SMs and/or DAPs to back-haul networks. In many situations where extremely short latencies and high availability are required (e.g., for fault isolation, failure restoration, critical control, etc.), high-speed wireline technologies such as fiber optics are the only option.

As opposed to wireline candidates, wireless communications generally offer connectivity for devices distributed in larger geographical areas at lower installation and maintenance costs since there is no need for cable running. However, it is very challenging to have high-speed and reliable communications using radio wave and the performance of wireless networks greatly changes depending on deployment scenarios. Therefore, the SGCN can use a the mix of both wireline and wireless technologies [3, 4]. Figure 3.1 summarizes and compares a number of representative wireless communications technologies from different perspectives including supporting data rate, coverage, mobility, indoor/outdoor use, power consumption and operating frequency band.

3.2 Short-Range Radios

3.2.1 IEEE 802.15.1/Bluetooth

Bluetooth (BT) was standardized as IEEE 802.15.1 but is currently managed by the Bluetooth Special Interest Group (SIG). This standard defines the physical (PHY) and medium access control (MAC) layers of the seven-layer Open Systems Interconnection (OSI) model for wireless communications over a short range (typically, in the order of tens of meters) and with low power consumption (from less than 1 mW up to hundreds of mW). BT is mainly designed for communications between closely located devices as a substitute for data transfer cables.

BT operates in the industrial, scientific and medical (ISM) radio band from 2400–2480 MHz and employs the frequency hopping spread spectrum (FHSS) communications technique. The transmitted data is divided into packets and each packet is transmitted on one of the designated channels. In its original release, BT 1.2 supported 1 Mbps data transfer. Bluetooth 2.0 with enhanced data rate version (EDR) and Bluetooth 3.0 with high-speed version (HS) increase the rate to 3 Mbps and higher. BT is a packet-based protocol with a master-slave structure. One master may communicate with up to seven slaves in a piconet. Packet exchange is based on the basic clock defined by the master. A connection has to be established to enable communications between a master and a slave. Devices not engaged in communications can enter one of several power and bandwidth-saving modes or drop the connection. Master and slave can switch roles, which may be necessary when a device wants to participate in more than one piconet.

3.2.2 Bluetooth Low Energy (BLE)

Recently, BT 4.0, also known as bluetooth low energy (BLE), has been introduced for ultra-lower-power, low-latency, low-data-rate and short-range communications. BLE aims to facilitate a wide range of applications with smaller form factor devices. By optimizing the protocol design for low-duty-cycle types of applications, BLE is expected to consume a fraction of the power of classic BT. In many cases, products will be able to operate for more than a year on a button cell battery without recharging. The pairing protocol for BLE devices is also simplified and enhanced in such a way that the connection setup time is much faster than that of classic BT, i.e., only a few milliseconds compared to several seconds.

3.2.3   Near-Field Communication (NFC)

NFC is a set of very short-range radio transmission standards for smart phones and similar devices. Communication between two devices is established by either bringing them into close proximity or through their physical contact, usually no more than a few centimeters. Communication is also possible between an NFC device and an un-powered NFC chip, called a tag. NFC operates at 13.56 MHz on ISO/IEC 18000-3 air interface and at rates ranging from 106 to 424 Kbps. NFC standards cover communications protocols and data exchange formats, and are based on existing radio-frequency identification (RFID) standards including ISO/IEC 14443 and FeliCa. The standards include ISO/IEC 18092 and those defined by the NFC Forum, which was founded in 2004 by Nokia, Philips and Sony, and now has more than 160 members. The Forum also promotes NFC and certifies device compliance. Present and anticipated applications include contactless transactions, data exchange, and simplified setup of more complex communications such as WiFi and BT/BLE.

3.3   Low-Rate Wireless Personal Area Networks (LR-WPANs)

3.3.1   IEEE 802.15.4

This is a standard that specifies the PHY and MAC layers for low-cost and low-power wireless communications technology designed for monitoring and control applications for residential and industrial settings. The PHY layer uses direct sequence spread spectrum (DSSS) modulation. It is highly tolerant of noise and interference and offers coding gain to improve link reliability. Standard binary phase-shift keying (BPSK) is used in the two low-speed versions (20 and 40 kbps operating on 868–868.6 and 915–928 MHz frequency bands, respectively), while offset-quadrature phase-shift keying (O-QPSK) is used for the higher-data-rate version (250 kbps operating on 2.4–2.4835 GHz frequency band). O-QPSK allows for more efficient non-linear power amplification for low power consumption. IEEE 802.15.4 uses carrier sense multiple access with collision avoidance (CSMA-CA) so that multiple nodes can share the same channel.

This standard is for low power consumption and thus may result in very long battery life for devices with low duty cycles. Besides, IEEE 802.15.4 chips are inexpensive and have small form factors. However, using IEEE 802.15.4, the typical radio communications range is only 10–75 m and nodes can only communicate in two simple topologies: point-to-point or multiple-point-to-point (star). Therefore, the coverage and flexibility of IEEE 802.15.4-based networks are quite limited. For more sophisticated topologies, a network layer that enables data routing between nodes is required. ZigBee, wireless highway addressable remote transducer protocol (WirelessHART) and IPv6 over low-power wireless personal area network (6LoW-PAN) are popular technologies that supply network and higher layers running on top of IEEE 802.15.4 PHY and MAC layers.

3.3.2 ZigBee

The most widely deployed enhancement to the IEEE 802.15.4 standard is ZigBee, specified by the ZigBee Alliance. It includes authentication with valid nodes, encryption for security, and a data routing and forwarding capability that enables mesh networking. The most popular use of ZigBee is wireless sensor networks (WSNs) using mesh topology. The main benefit of mesh topology is that any node can communicate with any other node, even in cases where they are not within range of each other, by relaying the transmission through multiple additional nodes. The network can thus spread out over a larger area. Furthermore, it increases network reliability as it still functions even if some nodes are disabled. There are usually alternate paths through the network to sustain the connection.

One of the key benefits of ZigBee is the availability of pre-developed applications. These upper-layer software additions implement specialized uses for ZigBee. Some of these applications include building automation for commercial monitoring and control of facilities, remote control, smart energy for home energy monitoring, home automation for control of smart homes, etc.

3.3.3 IEEE 802.15.4g

The IEEE 802.15.4g standard, also known as smart utility network (SUN), specifies amendments to the IEEE 802.15.4 standard to facilitate very large scale process control applications capable of supporting large, geographically diverse networks with minimal infrastructure and potentially millions of fixed end-points. It is principally designed for outdoor communications that operate in any of the regionally available license-exempt frequency bands, such as 700 MHz to 1 GHz, and the 2.4 GHz ISM band. Its supportable data rate varies from 5 to 400 kbps. As opposed to the IEEE 802.15.4/ZigBee standard, SUN employs time division multiple access (TDMA) based frequency-hopping MAC, namely coordinated sampled listening (CSL). The CSL allows receiving devices to periodically sample the channel(s) for incoming transmissions at low duty cycles. The receiving and transmitting devices are coordinated to reduce transmit overhead. SUN can support mesh topology and in fact, typical deployment scenarios for actual SUN implementations are based on this topology due to the fact that mesh networking can offer reliable access to/from the meters at a reasonable deployment cost. However, the detailed specification of the network topologies and routing protocols are out-of-scope of the IEEE 802.15.4g standard.

3.3.4 WirelessHART

This is a WSN technology based on HART and IEEE 802.15.4 compatible radios operating in the 2.4 GHz ISM radio band. It utilizes a time synchronized, self-organizing, and self-healing mesh architecture for process automation applications. Each WirelessHART network includes three main elements: Wireless field devices connected to process or plant equipment, gateways enabling communications between these devices and host applications connected to a high-speed backbone or other existing plant communications network. A network manager is responsible for configuring the network, scheduling communications between devices, managing message routes, and monitoring network health. The manager can be integrated into the gateway, host application, or process automation controller. WirelessHART is primarily designed to support industrial monitoring and control applications. Some of these applications include equipment and process monitoring, environmental monitoring, energy management, regulatory compliance, asset management, predictive maintenance, advanced diagnostics and closed-loop control.

3.3.5 6LoWPAN

In order to comply with the current and future standards and to ensure accessibility of WSNs, 6LoWPAN is introduced to resolve the inter-operability problem when integrating WSNs to the Internet. It defines encapsulation and header compression mechanisms that allow IPv6 packets to be sent to and received over IEEE 802.15.4 based networks. This protocol enables all the capabilities of IPv6 on very constrained devices and thus paves the road for the Internet of things. The data payload of IEEE 802.15.4 is 127 bytes while the standard IPv6 packet header is 40 bytes. To minimize the overhead, the adoption of IPv6 to small sensor devices starts by compressing the long IPv6 header, taking into account link-local information. The routing is adapted to the hop-by-hop "meshed" point of view of WSNs. Main routing functionalities are placed at the border routers rather than at sensor nodes having limited computation and memory capacities. Although the IP support in WSNs is nowadays a reality, large-scale application of 6LoWPAN in real world scenarios still remains limited.

3.3.6 Z-Wave

Z-Wave communicates using a low-power wireless technology designed specifically for remote control applications. The Z-Wave wireless protocol is optimized for reliable, low-latency communications of small data packets, unlike WiFi and other IEEE 802.11-based wireless local-area networks (LANs) that are designed primarily for high-bandwidth data flow. Z-Wave operates in the sub-gigahertz frequency range, around 900 MHz. This band competes with some cordless telephones and other consumer electronics devices, but avoids interference with WiFi and other systems that operate in the crowded 2.4 GHz band. The throughput is 40 kbps (9600 bps using old chips) and suitable for control and sensor applications. Each Z-Wave network may include up to 232 nodes and consists of two sets of nodes: controllers and slave devices. Nodes may be configured to retransmit the message in order to guarantee connectivity in multipath residential environments. Communication distance between two nodes is up to 30 m, and with up to a maximum of four hops between nodes, it gives enough coverage for most residential houses. Z-Wave is designed for embedding ease in consumer electronics products, including battery operated devices such as remote controls, smoke alarms and security sensors.

3.4 Wireless High-Speed LANs

3.4.1 IEEE 802.11 Infrastructured WiFi

WiFi is the most popular and successful technology for wireless local access. It enables a very wide range of portable and stationary consumer electronics devices with the ability to exchange data over a computer network, including high-speed Internet connections. It is based on IEEE 802.11 standards that specify over-the-air interfaces between a wireless client and an access point (AP) or between two wireless clients. While IEEE 802.15.1 and 802.15.4 standards are oriented towards low-power and low-data rate applications, IEEE 802.11 standard is primarily designed for broadband connections, as an extension of, or substitution for cabled LANs.

There are several specifications in the 802.11 family as follows. The original IEEE 802.11 provides 1 or 2 Mbps transmission in the 2.4 GHz band using either FHSS or DSSS. 802.11a is an extension to 802.11 that can provide up to 54 Mbps in the 5 GHz band. It uses an orthogonal frequency-division multiplexing (OFDM) encoding scheme rather than FHSS or DSSS. 802.11b is another extension to 802.11 that can support 11 Mbps transmission (with a fallback to 5.5, 2 and 1 Mbps) in the 2.4 GHz band, using DSSS. The term "WiFi" commonly refers to this specification. 802.11e adds quality-of-service (QoS) features and multimedia support to the existing 802.11b and 802.11a wireless standards, while maintaining full backward compatibility with these standards. 802.11g can support up to

54 Mbps in the 2.4 GHz bands. 802.11n employs multiple-input multiple-output (MIMO) for increased data throughput through spatial multiplexing and exploits spatial diversity through coding schemes for increased range. The PHY data rate can be as high as 250 Mbps (4–5 times faster than 802.11g). 802.11ac is built upon previous 802.11 standards, particularly the 802.11n standard, to deliver data rates of 433 Mbps per spatial stream, or 1.3 Gbps in a three-antenna design. 802.11ac operates only in the 5 GHz band and features support for wider channels (80 and 160 MHz) and beamforming capabilities by default to help achieve its higher wireless speeds. 802.11ad is a wireless specification under development that will operate in the 60 GHz frequency band and offer much higher transfer rates than previous 802.11 standards, with a theoretical maximum transfer rate of up to 7 Gbps.

The 802.11 standard defines two operating modes. In the infrastructure mode, wireless clients are directly connected to an AP in the basic star topology. APs are connected to a distribution network, usually wireline LANs, for Internet connection. In the ad-hoc mode, clients are connected to one another without any AP. The infrastructure mode is commonly used in most WiFi networks where wireline distribution networks for APs are available. In this mode, clients and an AP form a cell and a mobile client can roam from one cell to another. Since the communications range of each AP is short, typically from 30 to 100 m (depending on the radio specifications, operating frequency bands and surrounding environments), infrastructure WiFi networks have limited coverage. As a result, in some cases when extensive coverage is required, WiFi networks can be configured to work in a mesh topology.

3.4.2 IEEE 802.11 WiFi Mesh

In areas with difficult or expensive deployment of wired WiFi APs for Internet connection or in scenarios where a wireless ad-hoc network needs to be deployed within a very short time frame, WiFi mesh is a good solution. This kind of network configuration uses WiFi devices operating in the ad-hoc mode with mesh routing capability. WiFi mesh consists of mesh clients, mesh routers and gateways. Each client and router connects to several neighboring nodes and to a mesh gateway that aggregates mesh network traffic and routes it to the Internet. Nodes route data traffic among themselves over optimal paths and around failed or congested nodes. WiFi mesh networking address interference and reliability issues by separating access and back-haul on different frequencies: one 802.11b/g radio in 2.4 GHz band for access/client connections and one separate, dedicated 802.11a radio in 5.8 GHz band for back-haul transmission. For public safety applications, 4.9 GHz radio is used for both access and back-haul solutions.

3.5 Wireless MANs/WANs

3.5.1 IEEE 802.11 Municipal/City-Wide WiFi Mesh

WiFi mesh can be deployed to extend indoor/outdoor wireless networks over large geographical areas, e.g., municipality or city. In fact, it has been a widely used large-scale back-haul mesh networking solution. WiFi mesh can be deployed almost anywhere without the cost and disruption of running cabling or fiber. Meshes are resilient and require low maintenance. A well-designed WiFi mesh network automatically discovers the best route through the network and operates smoothly even if a mesh link or a node fails. Because the network is self-forming and self-healing, administration and maintenance costs are lower. In addition, a wireless mesh overcomes the line-of-sight issues that may occur when the deployment region is crowded with buildings or industrial equipment.

Existing city-wide deployments of WiFi networks demonstrate the clear applicability of WiFi as a wireless WAN. Minneapolis is just one example of a metropolitan installation in which WiFi is used not only for neighborhood network access but in WAN's back-haul portion of the system as well. A key advantage of WiFi for the WAN and back-haul is its use of a free, unlicensed spectrum. This makes it practical for a city to own and operate a large private wireless network for various applications. Other technologies, e.g., cellular networks, can provide the required service, but are usually owned and operated by large carriers who pay for the frequency licenses.

3.5.2 Cellular Networks

Cellular communications is one of the most successful and widely-used wireless communications technologies. It has experienced exponential growth in the past two decades with global service coverage and billions of users. Although, cellular communications was originally developed for voice services, i.e., allowing a person with a cell phone to make or receive a call from almost anywhere even while on the move, it has been evolving and nowadays is extensively used for a wide range of data services. Most cellular systems operate in the licensed frequency spectrum of 824–894/1900 MHz. Data transmission rate of this technology varies from a few tens of kbps to hundreds of Mbps, depending on technologies, distances and deployment scenarios.

The cellular network has gone through four generations. The first generation (1G) is analog in nature. It uses frequency division multiple access (FDMA) and analog frequency modulation. Digital technologies have been introduced in the second generation (2G) to improve voice quality and provide a set of rich voice features. Time-division multiple access (TDMA) and code-division multiple access (CDMA) are also used to increase the network capacity. The third generation (3G) integrates cellular phones into the Internet world by providing high-speed packet-switching data transmission in addition to circuit-switching voice transmission. The fourth generation (4G) refers to all-Internet Protocol (IP) packet-switched networks, giving ultra mobile broadband access, e.g., the long-term evolution (LTE).

LTE has introduced a number of new technologies when compared to the 2G/3G cellular systems. They enable LTE to operate more efficiently with respect to spectrum use, and also provide the required higher data rates. First, OFDM technology has been incorporated into LTE to enable efficient broadband transmission while still providing a high degree of resilience to reflections and interference. Second, orthogonal frequency division multiple access (OFDMA) is used in the downlink while single-carrier frequency-division multiple access (SC-FDMA) is used in the uplink. Third, one of the main problems of previous cellular systems is that of multiple signals arising from many reflections that are encountered. By using MIMO, these additional signal paths can be used advantageously to increase the throughput. Finally, with the very high data rate and low latency requirements for next-generation cellular systems, LTE system architecture evolution (SAE) modifies the system architecture for better performance. One change is that a number of the functions previously handled by the core network have been transferred out to the periphery. Essentially, this provides a much "flatter" form of network architecture. In this way, data can be routed more directly to its destination and thus latency can be reduced. LTE can achieve 100 and 50 Mbps for peak downlink and uplink data rates, respectively. Latency of round trip packet transmission is around 10 ms.

LTE-Advanced (LTE-A) has been developed with similar technology evolutions as LTE, e.g., OFDM, OFDMA, SC-FDMA, MIMO, etc. For LTE-A, the use of MIMO evolves further and more advanced techniques with additional antennas enable the use of additional paths. Compared to LTE, LTE-A exhibits a number of advantages. Its peak data rates for downlink and uplink are 1 Gbps and 500 Mbps, respectively. Its spectrum efficiency is three times greater than that of LTE. LTE-A is capable of supporting scalable bandwidth use and spectrum aggregation where non-contiguous spectrums are used. The latency is less than 5 ms one way for individual packet transmission.

It is noted that LTE is not the only candidate 4G technology. Mobile worldwide inter-operability for microwave access (WiMAX) is also promising since it can offer very high data rates and high levels of mobility. However, it seems less likely that WiMAX will be adopted as compared to LTE. One of the main reasons why LTE appears better positioned for adoption is that it has been designed to be well integrated with other cellular technologies, making for smoother and more economical transitions from 3G to 4G.

3.5.3   IEEE 802.16/WiMAX

WiMAX technology is part of the IEEE 802.16 series of standards for wireless MAN. The main objective of WiMAX is to achieve worldwide inter-operability for microwave access. The original WiMAX standard (IEEE 802.16) proposes the usage of 10–66 GHz frequency spectrum. The WiMAX forum has published a subset of the range for inter-operability. For fixed communications 3.5 (licensed) and 5.8 GHz (unlicensed) bands have been dedicated, while for mobile communications frequency bands 2.3, 2.5 and 3.5 GHz (licensed) have been assigned.

For last mile access, WiMAX operates similarly to WiFi but at much higher data rates, over greater distances and for a greater number of users. It is designed to provide 30–40 Mbps data rates (and up to 1 Gbps for fixed stations) and has a range of up to 50 km. WiMAX also has benefits over WiFi in terms of connection quality. When multiple users are connected to a WiFi AP, they contend with each other for channel access using contention-based CSMA mechanism, and thus they can experience varying levels of bandwidth. WiMAX employs a MAC layer based on a grant-request mechanism to authorize the exchange of data. This feature allows better exploitation of the radio resources, in particular with smart antennas, and independent management of user traffic.

Beyond just providing broadband last mile access, WiMAX can be used for creating wide-area wireless back-haul network that can operate in point-to-point or mesh modes. When backhaul-based WiMAX is deployed in mesh mode, it does not only increase wireless coverage, but it also provides features such as lower back-haul deployment cost, rapid deployment, and re-configurability. Various deployment scenarios include city-wide wireless coverage, back-haul for connecting 3G radio network controller (RNC) with base stations, etc.

In addition to fixed WiMAX, mobile WiMAX enables cell phone-like applications. For example, mobile WiMAX enables streaming video to be broadcast from a speeding police or other emergency vehicle at high speeds. Mobile WiMAX supports optimized handover schemes with latencies of less than 50 ms to ensure real-time applications without service degradation. Flexible key management schemes assure that security is maintained during handover. It can potentially replace cell phones and mobile data offerings from cell phone operators such as High-Speed Downlink Packet Access (HSDPA). However, with the current ramp-up in LTE deployments, the window of opportunity for WiMAX is closing rapidly.

3.5.4 Satellite Communications

Satellite communications provides a long-distance microwave radio relay technology complementary to that of communication cables. They are also used for mobile applications such as communications to ships, vehicles, planes and hand-held terminals, and for televisions and radio broadcasting. Dependent upon the type of satellite communications system and the orbits used, it is possible to provide complete global coverage. As a result, satellite communications systems are used for providing communications capabilities in many remote areas where other technologies would not be viable. This technology can support data rates ranging from 9.6 Kbps to a few tens of Mbps. However, cost and communication

Table 3.1 Representative wireless technologies for SGCN implementation

Advantages	Disadvantages	Applicability
Zigbee (IEEE 802.15.4, ZigBee Alliance) and SUN (IEEE 802.15.4g)		
Low power consumption; Very low cost; Self-organizing, secure, and reliable mesh network; Network can support a large number of users; SEP for HANs is available	Very short range; Does not penetrate structures well; Low data rates; Interfered by other devices operating in the same radio frequency band	HAN (Zigbee) NAN (IEEE 802.15.4 and 802.15.4g)
Wi-Fi (IEEE 802.11b/g/n) and **WiFi Mesh**		
Low-cost chip sets; Widespread use and expertise; Deployed in most residential/commercial buildings; Stable and mature standards	Does not penetrate cement buildings or basements; Security issues with multiple networks operating in same locations	HAN (infrastructured) NAN (mesh) WAN (mesh)
3G Cellular (CDMA2000, EDGE, UMTS, HSPA+)		
Expensive infrastructure already widely deployed, stable and mature; Well standardized; Equipment prices keep dropping; Readily available expertise in deployments; Cellular chipset very inexpensive; Large selection of vendors and service providers	Utility must rent the infrastructure from a cellular carrier for a monthly access fee; Technology is in the transition phase to LTE deployment; Public cellular networks not sufficiently stable/secure for mission critical/utility applications; Not well-suited for large data/high bandwidth applications	WAN NAN (in some cases)
4G Cellular (LTE)		
Low latency, high capacity; Fully integrated with 3GPP, compatible with earlier 3GPP releases; Full mobility for enhanced multimedia services; Carrier preferred protocol; Low power consumption	Utility must rent the infrastructure from a cellular carrier for a monthly access fee; Not readily available in many markets/still in testing phases in others; Equipment cost high; Vendor differentiation still unclear; Lack of expertise in designing LTE networks; Utilities' access to spectrum	WAN NAN (in some cases)
WiMAX (IEEE 802.16)		
Efficient backhaul of data (aggregating 100's access points); QoS supports service assurance; Battery-backup improves reliability and security; Simple, scalable network rollout and Customer-Premises Equipment (CPE) attachment; Faster speeds than 3G cellular; Large variety of CPE and gateway/base station designs	Limited access to spectrum licenses in the US; Trade-off between higher bit rates over longer distances; Asymmetrical up and down link speeds; User shared bandwidth; Competing against future 4G cellular standards for high-capacity, all-IP networks	WAN NAN (in some cases)

latency are the two key disadvantages of this technology. It is very expensive to build, launch and maintain satellites. This means that the operational costs are high, and therefore, the cost of renting or buying space on the satellite is very costly. In addition to these costs, the user needs a specialized terminal to communicate with the satellite. Moreover, as distances are much greater than those involved with terrestrial systems, propagation delay can be an issue, especially for satellites using geostationary orbits. The round trip from the ground to a geostationary satellite and back is approximately 250 ms.

3.6 Candidate Wireless Technologies for SGCN Implementation

As presented in the previous section, there are many different communications technologies that could be used to implement the SGCN. A preferred standard would be the one that is compatible or common across HAN, NAN and WAN domains since this would lower the equipment cost and simplify the implementation. Recently, the introduction of the smart energy profile (SEP) has promoted IEEE 802.15.4/ZigBee as the most preferred standard for HAN applications. If this technology is also considered for the implementation of NAN, the same radio could be used in the devices installed at homes and utilities. However, using the same or similar technology for all segments of the SGCN is very challenging due to the fact that those segments differ significantly in terms of propagation channel characteristics, required radio communications range, power consumption, network or device lifetime, network scale, etc. Each network also needs to carry different volumes and types of traffic with different QoS requirements. Therefore, different segments of the SGCN may need different communications technologies. Furthermore, there could be more than one technology in each segment. Table 3.1 gives a brief overview on strengths and weaknesses of a number of wireless technologies and SGCN segments that they may be applicable for. More details on the technology selection are discussed in the following subsections.

3.6.1 Wireless Technologies for HANs

For this network segment, IEEE 802.15.4/ZigBee has appeared to be a global standard for home energy management, demand response and other smart home applications. In fact, many world leading utilities, energy service providers, product manufacturers and technology companies are supporting the development of ZigBee SEP. Several other standard organizations are also involved with extending the reach of ZigBee SEP to more homes around the world. Most of the SMs commercially available today are integrated with ZigBee radio and SEP (in addition to their primary interface for NAN communications). A typical example is the Focus AX/SD meter manufactured by Landis+Gyr (Toshiba) [5]. This kind of meter employs ZigBee and SEP so that it can communicate with ZigBee-based HAN and work as a HAN gateway.

Another candidate technology that could be promising for HANs is IEEE 802.11 WiFi due to the fact that it is the most widely used wireless LAN technology and can support relatively high data rates at reasonably low costs. More than two billion WiFi-certified devices have been installed for home and industrial applications. WiFi is based on very mature technology with well-proven encryption, authentication and end-to-end network security. It also has a mature ecosystem and widely-demonstrated inter-operability. The feasibility of WiFi for HANs is even more convincing with the introduction of low-power WiFi standards for IP smart objects. Low-power WiFi devices have the advantages of native IP-network compatibility and well-known protocols and management tools.

Additionally, BT and BLE technologies are also considered for the implementation of short-range, low-power and low-data-rate communications between home energy management server (HEMS) and home displays/control panels that allow consumers to track their electricity energy usage and to customize/optimize their energy consumption profiles. Specifically, BLE can enable consumers to interact with their HEMS using their smart phones and/or tablets. NFC may be used to simplify the setup of more complex communications such as WiFi and BT/BLE that are deployed in HANs.

3.6.2   Wireless Technologies for NANs

IEEE 802.15.4g/SUN radio standard, a global wireless networking standard aiming to enable inter-operable communications between smart grid devices, is likely to be the most widely adopted technology for NANs. Its last-updated standard represents a huge leap forward in establishing common and consistent communication specifications for utilities deploying smart grid technologies. A large number of technology companies and manufacturers are driving for the adoption of IEEE 802.15.4g/SUN radio for the implementation of smart metering products. These companies include Landis+Gyr (Toshiba) [5], Itron [6], Silver Spring Networks [7], Trilliant [8], etc.

IEEE 802.11 WiFi mesh is also a competitive technology for this network segment, especially when broadband data rates need to be provisioned for emerging SG applications such as advanced distribution automation, power quality monitoring, and so on. Municipal-scale WiFi network infrastructure has already been deployed using 802.11 technology. This includes systems, for example, that provide access covering up to 500 m from the access point, interconnected by point-to-point links based on 802.11 technology and using proprietary mesh protocols. Modern municipal WiFi networks typically also support 4.9 GHz access for public safety networks that are also based on 802.11 technology. Newer developments in the 802.11n standard, including support for transmit beam-forming, may further enhance the use of WiFi for these outdoor applications. Existing municipal 802.11-based networks are the appropriate scale for NAN. WiFi can connect hundreds of devices on buildings and pole tops in a variety of terrains. The 4.9 GHz public safety application shows that WiFi can be re-banded to support lightly licensed spectrum with different channel sizes. NANs might benefit from operating in lightly licensed spectrum similar to the 4.9 GHz spectrum that has been set aside for public safety applications in the United States. Additional enhancements for NAN can come from work being done within the IEEE 802.11s Task Group to standardize a mesh networking protocol.

It is noted that other technologies, e.g., cellular communications and WiMAX, can also be used for connecting each individual smart meters to WANs or utilities control centers. Unfortunately, at this moment, these technologies might not be economically feasible since the utility would need to pay monthly charges for each of the millions of connected devices. Also, connecting a large number of devices using cellular or WiMAX may pose challenges with respect to resource scheduling and allocation. However, when technologies evolve, the communication cost per connected device tends to decrease while the network capacity increases and resource allocation becomes more sophisticated, cellular and WiMAX might become a promising candidate technology for NANs. Currently, cellular and WiMAX are selected to implement NANs in areas that are most prone to disasters like hurricanes and major storms. Besides, these two technologies and satellite communications are employed for NANs in rural, remote and low-density areas since they are the only viable solution for those regions.

3.6.3 Wireless Technologies for WANs

3G/4G cellular networks are the fastest and least-expensive way for electric utilities to deploy a WAN to monitor and control the SG. Cellular network operators can wirelessly connect DAPs (that aggregate traffic for NANs) and other important grid assets such as breakers, sensors, remote terminal units, transformers and substations directly to the utility's operation centers. This dramatically reduces the utility's up-front deployment costs and time-frames and leverages the cellular carriers' massive investments in network operations and maintenance to reduce the overall cost of ownership. WiMAX currently exhibits a higher bandwidth and a lower latency, compared to 3G cellular communications. However, with the imminent LTE deployment from multiple carriers, WiMAX might lose those advantages. Unlike WiMAX deployments, LTE will mostly reuse existing cellular networks and should be a straightforward evolution of the 3G cellular networks. Due to the popularity of cellular networks and the associated economies of scale that come with a large number of subscribers, 4G cellular is expected to be more economical and more widely available than WiMAX. It can be seen that the use of 3G/4G cellular for WANs has been promoted over the last few years. For example, in 2012, Itron purchased SmartSynch [9] which is a leading proponent of cellular-based communications in order to integrate cellular communications into its OpenWay wireless mesh NAN infrastructure. Likewise, Landis+Gyr has recently acquired smart grid startup Consert [10] who has been building a smart home service utilizing 3G networks in order to strengthen its WAN technology. Additionally, a number of manufacturers (e.g., Silver Spring Networks, one of the world's leading companies building wireless communications networks for utilities) have already embedded cellular communications modules into their smart metering network gateways or DAPs to make WAN connectivity readily available for fast deployment.

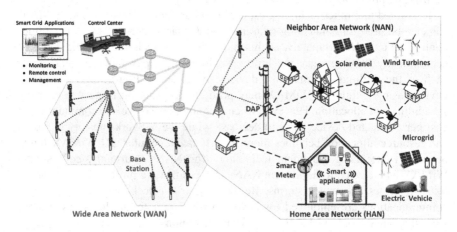

Fig. 3.2 The SGCN as an integration of various communications technologies

Besides cellular and WiMAX, WiFi is another candidate technology for WANs. Existing city-wide deployments of WiFi networks demonstrate the clear applicability of WiFi as a SG's WAN technology. Minneapolis is just one example of a metropolitan installation in which WiFi is used not only for neighborhood network access but in WAN backhaul portion of the system as well. Today, such metropolitan area WANs, incorporating standard 802.11 WiFi in point-to-point or point-to-multipoint links, embody a variety of proprietary network management approaches and demonstrate that WiFi technology could be similarly incorporated into the future standardized SG management framework for WAN communications. A key advantage of WiFi for WAN is its use of a free, unlicensed spectrum. This makes it practical for a city or utility to own and operate a large private wireless network for the SG. Cellular and WiMAX data networks can provide the required service, but are usually owned and operated by large carriers. S&C Electrics [11], Tropos Networks [12] and Fujitsu [13] are among a few companies that have been using WiFi radio for their WAN and back-haul communications products.

It is worth noting that, a wide range of wireline and wireless technologies could be used for the SGCN implementation, as mentioned in Sect. 3.1 of this chapter. DSL, leased line, PLC, or fiber optics should be exploited in scenarios where their advantages in costs, supportable data rates, and delays are justifiable.

For example, DSL or fiber optics are readily available in many apartments, buildings, and substations, therefore, they could be used for various smart metering, monitoring, and automation applications. Satellite communications seems to be the only viable solution to provide connectivity to sparsely-distributed and remote areas despite the fact that it suffers from latency and cost issues. In fact, most commercial SG communications modules and products manufactured with given primary communications technologies have serial and Ethernet ports supporting the connections with other communications interfaces using alternative technologies. Another important note is that there are regional differences in technology adoption. For example, WiMAX has been widely deployed in Australia while LTE and private wireless mesh utility networks are more favorable in North America. PLC is a dominant technology for access communications in Europe, however, it has a very limited deployment in North America.

3.7 A Typical Implementation of the SGCN

3.7.1 Network Architecture

As addressed in preceding section, there is no one-size-fit-all communications technology for the SGCN due to the diversity in service coverage area, population density, financial budget, as well as performance requirements. In order to illustrate the use of multiple technologies in the SGCN, Fig. 3.2 presents a widely-accepted network architecture. A low-power and low-data-rate radio technology provides the connectivity for devices in HANs and broadband wireless mesh networks connect SMs in NANs and relay information to DAPs that are back-hauled by cellular and wireline networks.

For HANs, with ZigBee SEP, IEEE 802.15.4/ZigBee can support numerous SG applications that monitor, control, and automate the delivery and consumption of electric energy. In fact, SEP was selected by the NIST as a standard profile for smart energy management in home devices. Recently, ZigBee IP has been introduced as the first standard-based IPv6 specification for wireless sensor networks. It enables Internet of Things (IoT) and M2M communications. The integration of ZigBee SEP and ZigBee IP allows smart devices in homes/buildings, PEVs, as well as microgrid subsystems immediately become available to the network. This facilitates more flexible and autonomous control applications.

The wireless mesh has been considered as the most promising solution for NANs since it has low deployment and maintenance costs as well as excellent resilience to node/link failures. A wireless mesh network can be established almost anywhere without the cost and disruption of running cabling or fiber. It provides a multi-path, multi-hop connectivity that is absolutely necessary for outdoor deployment. IEEE 802.11 WiFi radio is employed in this segment due to its maturity and low-cost while providing high-speed broadband communications (up to a few tens or even hundreds of Mbps) that emerging advanced SG applications are likely to require (as opposed to IEEE 802.15.4-based technologies that can only support up to a few hundreds of kbps).

Cellular networks have a very wide coverage while offering a high data rate and low latency. Especially, LTE technology can deliver low cost per bit, advanced radio resource management and scheduling, and high-performance connectivity needed to address the challenges of machine-to-machine (M2M) communications in the SG. This technology has been considered as a future-proof solution for DAPs and many other SG' devices that need to remain in the field for 10 or 20 years. The wireline backbone in Fig. 3.2 can be implemented by various broadband technologies such as leased lines or fiber optics that can provide gigabit-per-second and sub-millisecond connectivity.

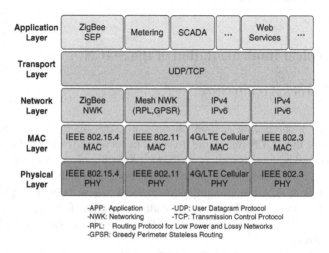

Fig. 3.3 Protocol stacks employed in the SGCN

3.7.2 Protocol Architecture

The three segments of the SGCN shown in Fig. 3.2 employ different communications technologies and thus operate with different protocol stacks. The gateways that interconnect them need to perform protocol conversion. Figure 3.3 depicts conceptual OSI protocol stacks of each segment. Since an end-to-end IP network architecture offers many distinguishing features facilitating open standards and inter-operability with scalability, stability, and cost efficiency, it is assumed in Fig. 3.3. IPv4/IPv6 is employed in the network layers of the cellular and wireless back-haul networks. For IEEE 802.15.4/ZigBee and IEEE 802.11 WiFi mesh networking of HAN and NAN segments, respectively, their network layers are implemented with their own routing protocols for reliability and power efficiency to address unique challenges in these segments. However, their MAC and network layers may include sub-layers for IP networking. For example, 6loWPAN (RFC 6282) sub-layer operates on top of the IEEE 802.15.4 MAC protocol to support the transmission of IPv6 packets over IEEE 802.15.4 networks. TCP/UDP protocol is used for data transport. ZigBee SEP and other application profiles are deployed to support SG applications.

References

1. C. W. Chao, Q. D. Ho, and T. Le-Ngoc, "Challenges of power line communications for advanced distribution automation in smart grid," in *Proc. the 2013 IEEE Power & Energy Society General Meeting*, Vancouver, Canada, 21–25 Jul. 2013.
2. Q. D. Ho, C. W. Chao, M. Derakhshani, and T. Le-Ngoc, "An analysis on throughput and feasibility of narrow-band power line communications in advanced distribution automation scenarios," in *Proc. the 2014 IEEE International Conference on Communications (ICC 2014)*, Sydney, Australia, 10–14 Jun. 2014.
3. Q. D. Ho, Y. Gao, and T. Le-Ngoc, "Challenges and research opportunities in wireless communications networks for smart grid," *IEEE Wireless Communications*, pp. 89–95, Jun. 2013.
4. Q. D. Ho and T. Le-Ngoc. *Smart grid communications networks: Wireless technologies, protocols, issues and standards*, Chapter 5 in Handbook on Green Information and Communication Systems (Editors: S. O. Mohammad, A. Alagan, and W. Isaac), Elsevier, Summer 2012.
5. "Landis+Gyr." [Online]. Available: http://www.landisgyr.com.
6. "Itron." [Online]. Available: http://www.itron.com.
7. "Silver Spring Networks." [Online]. Available: http://www.silverspringnet.com.
8. "Trilliant." [Online]. Available: http://www.trilliantinc.com.
9. "Smart synch." [Online]. Available: http://www.smartsynch.com.
10. "Consert." [Online]. Available: http://www.consert.com.
11. "S&C Electrics." [Online]. Available: http://www.sandc.com.
12. "Tropos Networks." [Online]. Available: http://www.tropos.com.
13. "Fujitsu." [Online]. Available: http://www.fujitsu.com.

Chapter 4
Wireless Routing Protocols for NANs

In order to determine suitable routing protocols for the SGCN, this chapter surveys various existing routing protocols designed for wireless mesh networks. The focus is given to protocols that could be used for the NAN which is the most important segment of the SGCN. Based on this survey, the main operating principles along with weaknesses and strengths of the two candidate routing protocols are presented. Specifically, the Greedy Perimeter Stateless Routing (GPSR) and the Routing Protocol for Low Power and Lossy Networks (RPL) are chosen to represent the location-based and self-organizing coordinate routing protocols, respectively. Finally, since the robustness is one of the key requirements of the NAN, this chapter presents a proactive parent switching (PPS) scheme, as an extension for RPL, to handle network element failures.

4.1 Routing in Wireless Networks

4.1.1 Wireless Routing Protocol Classification

Over the last few decades, several routing protocols have been proposed and studied for wireless ad-hoc and sensor networks [1–6]. They can be classified into different protocol families depending on underlying network structure (i.e., flat, hierarchical, and location-based routing) and protocol operation (i.e., multipath-based, query-based, negotiation-based, QoS-based, and coherent-based). Routing protocols can be proactive (i.e., each node actively collects current network status and maintains one or more tables containing routing information to every other nodes in the network) or reactive (i.e., routes are created when required by performing route discovery and selection procedures on-demand). In the following sections, key

© The Author(s) 2014
Q.-D. Ho et al., *Wireless Communications Networks for the Smart Grid*,
SpringerBriefs in Computer Science, DOI 10.1007/978-3-319-10347-1_4

Routing protocols for wireless networks

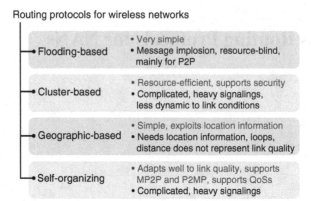

Fig. 4.1 Wireless routing protocol classification

features, advantages and disadvantages of a number of representative routing protocols (in Fig. 4.1) are investigated in order to facilitate the selection of candidate protocols for SG's NANs.

4.1.2   Flooding-Based Protocols

Flooding-based protocols enable point-to-point (P2P) traffic patterns and rely on broadcasting data and control packets by each node into the entire network. In its conventional implementation, a source node sends a packet to all of its neighbors, each of which relays the packet to their neighbors, until all the nodes in the network (including the destination) have received the packet. Despite its simplicity, pure flooding suffers from many disadvantages including implosion, i.e., redundant copies of messages are sent to the same node by different neighbors or through different paths, and resource blindness, i.e., flooding lacks consideration for energy constraints of nodes when transmitting packets [7]. Flooding protocols are only particularly useful for P2P communications among a small number of mobile nodes without the need for any routing algorithm and topology maintenance. Those disadvantages render the application of flooding-based protocols to the NAN infeasible.

4.1.3   Cluster-Based Protocols

Cluster-based protocols are based on a hierarchical network organization. Nodes are grouped into clusters, with a cluster head elected for each one. Data transmission typically goes from cluster members to the cluster head, before going from the cluster head to the sink node. Since cluster heads are responsible for relaying and processing high volume of data, they typically have higher energy and computation capability. This kind of routing can support multi-point-to-point (MP2P), point-to-multi-point (P2MP) and P2P traffic. Clusters are built and maintained as a function of the various parameters of the nodes and system. Those parameters include node energy, link quality, traffic pattern, data correlations between nodes, etc. [8–14]. The drawbacks of this class of routing protocols are that it cannot capture the link dynamics, and head selection, cluster formation and maintenance introduce significant signaling overhead. Besides, protocols like LEACH [13] and HEED [11] assume that TDMA and CDMA are used for intra-cluster and inter-cluster communications, respectively, and that nodes can tune their communication range through transmission power. Those assumptions make them nearly impractical for real deployment. Fortunately, splitting the network into smaller clusters efficiently limits the data flooding area. This offers benefits in scalability, lifetime, and energy consumption. Additionally, since nodes physically close to each other are likely to sense correlated events, data can be efficiently aggregated at the cluster head to reduce network load. Implementation of security is also easier since cluster heads can act as trusted entities in the network. These advantages seem to be very attractive to the NAN. First, the NAN is naturally organized into multiple clusters. Each cluster serves a few thousands of SMs and data is managed by the DAP acting as the cluster head. Second, SMs that have some underlying correlation can be placed within the same cluster. For example, SMs located in the same distribution feeder may send similar notification messages at the same time when their feeder fails. If those messages are gathered by their cluster head, redundancy can be detected and resolved efficiently to minimize network traffic volume while still assuring that no important information is lost throughout the network. Finally, the heads of each cluster can offer important security features that are required by the NAN.

4.1.4 Geographic Protocols

Geographic protocols (GEO) route the traffic based on the location knowledge of a node, its neighbors and the sink node. Greedy forwarding (GF) is the simplest form of GEO. When a node receives a message, it relays the message to its neighbor geographically closest to the sink [15]. Since geographic distance is not necessarily radio communication distance, the drawback of GF is that the selection of next hop merely based on geographic distance may lead to void areas where the traffic cannot advance further towards the destination. More advanced location-based routing protocols that attempt to improve the delivery rate are proposed in [16–20]. The advantage of GEO is that it can achieve network wide routing while maintaining only neighborhood information at each node, hence significantly reducing signaling overheads and the complexity of the routing solution. This kind of routing protocol is applicable in the case where geographical locations of each node are known in advance.

4.1.5   Self-Organizing Coordinate Protocols

Self-organizing coordinate protocols counteract the biggest drawback of GEO by building a viable coordinate system based on communication distance rather than geographic distance. The aim of such coordinate systems in the context routing protocols is not to mimic geographic location but rather to be of use for feasible routing solutions. The routing protocol for low-power and lossy networks (RPL) is a representative protocol that captures most of the ideas introduced by self-organizing coordinate protocols [3]. Advantages of RPL can be summarized as follows. First, RPL basically constructs a directed acyclic graph (DAG) whose structure matches the physical structure of NAN. Root nodes represent DAPs, leaf nodes represent SMs, and other nodes inside the DAG represent routers that maintain connectivity between root and leaf nodes. Second, MP2P and P2MP, typically required by the NAN, are the primary communications supported by RPL over its DAG. Third, by employing different routing metrics and cost functions, RPL can construct multiple instances of DAGs over a given physical network. Each instance can be dedicated for a specific routing objective. This facilitates QoS differentiation and provisioning for different types of traffic that NANs need to carry. Moreover, with the trickle timer that governs the network state update, RPL requires less signaling overhead and thus it is more energy-efficient. Finally, the roots of DAGs can act as trusted entities that enable security in the network.

4.2   Routing in NANs

NAN is an important constitutive segment of the SGCN since it provides the connectivity between SMs and the utilities to enable various key SG applications. On one hand, it can be considered as an outdoor wireless sensor and actuator network (WSAN). On the other hand, it exhibits many characteristics and challenges that are not found in a general outdoor WSAN. Most of NAN devices are installed in a close proximity to power lines and equipment. For example, SMs are typically in distribution feeders before the electricity is delivered into the apartments or buildings while smart sensors and controllers are along power lines, on utility poles, or in substation areas. Therefore, the communications of these devices can be affected by ambient electro-magnetic interference (EMI) produced by power lines and power switching equipment. Such EMI results in impulsive noise that is distinguished from the thermal Gaussian noise produced in the receivers of the devices themselves. Besides, NAN devices are usually powered by the grid itself; however, in cases of power outages in the grid, they still have to stay operational by switching to their battery power supplies. Many other distinguishing characteristics and requirements of the NAN are similar to those that have been addressed by the SGCN in Sect. 1.3.1. The NAN connects millions of SMs, routers, and gateways that are distributed over a vast geographical area. These devices are mostly installed outdoor and operate in hash environments (low/high temperatures, obstacles, rain, snow and so on). This network therefore needs to be scalable, self-organizing, and robust. Cyber security requirements are also critical since the NAN conveys a huge volume of private information of residential/commercial consumers (e.g., identities, power consumption profiles and habits, real-time residential/business activities, etc.) as well as vital sensor and control signals necessary to harmonize/optimize the operations of the power grid and to reduce energy consumption. Besides, traffic patterns and QoS requirements for NANs are different from those of conventional applications. For example, periodic meter reading requires high reliability but can tolerate latency and jitter. However, emergency messages (e.g., in case of power grid failures) are generated randomly in bursts but require very stringent latency. Therefore, developing routing protocols for NANs is a challenging task. The authors in [21–23] give a survey of routing protocols selected for NAN scenarios using various communications technologies and networking. This work, as previously discussed, focuses on wireless mesh networks due to their advantages in deployment, operation and maintenance costs.

Greedy perimeter stateless routing (GPSR), a well-known implementation of GEO, and the routing protocol for low-power and lossy networks (RPL) have been identified as the most promising routing protocol for NANs. One of the compelling advantages of GPSR is that it can achieve network wide routing while maintaining only neighborhood information at each node. The simplicity of GPSR leads to good scalability since it is no necessary to keep routing tables up-to-date and to have a global view of the network topology and its changes. GPSR protocols allow routers to be nearly stateless because forwarding decisions are based on location information of the destinations and the location information of one-hop neighbors. No routing table is needed to be constructed or maintained. A new node can join the network easily by locally exchanging information with existing nodes in its vicinity. Since establishment and maintenance of routes are not required, signaling overhead and computational complexity of GPSR can be kept at a considerably low level. In addition to these advantages, the fact that locations of NAN devices are fixed and accurately known promotes GPSR as a promising solution for NANs.

RPL is a representative protocol which captures most of the ideas introduced by self-organizing coordinate protocols. RPL exhibits many advantages that are desirable in the NAN setting. First, the tree-like structure constructed by RPL matches well with the physical deployment and communication model of SMs (nodes) and DAPs (sinks or roots). Second, RPL is designed to be able to incorporate various types of routing metrics and constraints that can be addictive, multiplicative, inclusive, exclusive and so on. Therefore, both QoS-aware and constraint-based routing disciplines can be supported. Third, RPL allows multiple logical routing graphs to operate concurrently and independently to provide QoS differentiation for different classes of traffic in the NAN.

Performance of a geographic routing protocol in realistic smart metering scenarios is presented in [24]. Using simulations, received packet ratios given by the protocol are measured against network scales, offered traffic rates, and placements of routers and DAPs. For the small-scale scenario (350 SMs, 2 routers and 1 DAP), the simulations show that the system performs with a received packet ratio of 100 % for a message frequency of 1 message per 4 h. However, success rate decreases with increasing message frequency due to collisions in some central nodes. For a large-scale scenario (17,181 SMs, multiple routers and DAPs), an overall success rate of 99.99 % for a message frequency of 1 message per 4 h is observed. In this case, it is noted that there are some isolated zones due to coverage gaps. Geographical distributions of packet success rate and hop-count are analyzed in order to determine the suitable number and placement of routers and DAPs that would result in an improved performance. The authors in [25, 26] present extensive study on transmission reliability and latency of GPSR against channel conditions, network scales, and per-meter traffic in practical NAN scenarios specified by SG standards. The results demonstrate the effects of realistic channel models and increasing network loads introduced by emerging SG applications to the system performance of NANs.

Performance of RPL implemented in an experimental platform using TinyOS is presented in [27]. The results in [27] indicate that RPL performs similarly to the Collection Tree Protocol (CTP), the de-facto standard data collection protocol for TinyOS 2.x [28], in terms of packet delivery and protocol overhead. Compared to CTP, RPL can provide additional functionalities, i.e., it is able to establish bi-directional routes and support various types of traffic patterns including MP2P, P2MP and P2P. Therefore, the authors in [27] conclude that RPL is more attractive for practical wireless sensing systems.

The work in [29] analyzes the stability of RPL whose DAG is built based on link layer delays. It is observed that the fluctuation in the delay introduced by the IEEE 802.15.4 MAC layer negatively influences RPL's stability. That fluctuation forces the nodes to change their best parent along the routing path so frequently that it results in significant end-to-end delay jitters. In order to dampen the link layer delay fluctuation, the author proposes the use of memory in delay calculation. Simulations of a small network demonstrate that the proposed solution can reduce the mean and variance of the end-to-end latency and thus improve the protocol stability.

The authors in [30] provide a practical implementation of RPL with a number of proper modifications so as to fit into the AMI structure and meet stringent requirements enforced by the AMI. In particular, expected transmission time (ETX) link metric and a novel ETX-based rank computation method are used to construct and maintain the DAG. ETX is measured by a low-cost scheme based on a MAC layer feedback mechanism. A reverse path recording mechanism to establish the routes for downlink communications (i.e., from gateways to end-devices) is also proposed. The mechanism is purely based on the processing of uplink unicast data traffic (i.e., from end-devices to gateways), and hence does not produce extra protocol overhead. Extensive simulation results in [30] show that, in a typical NAN with 1,000 SMs, and in the presence of shadow fading, the proposed RPL-based routing protocol outperforms some existing routing protocols like ad-hoc on demand vector routing (AODV), and produces satisfactory performances in terms of packet delivery ratio and end-to-end delay.

Self-organizing and self-healing solutions for RPL are proposed in [31]. SMs are able to automatically discover DAPs in their vicinity and setup a single or multi-hop link to a selected DAP. A distinguishing feature in [31] is that DAPs may choose to operate at different frequencies in order to accommodate a scalable large network consisting of multiple trees. SMs perform channel scanning to detect DAPs and select the best one. Also, SMs can detect loss of connectivity arising from failed nodes/links/concentrator and automatically recover from such failures by dynamically connecting to an alternative concentrator in their vicinity. Numerous performance parameters of the proposed RPL are studied by simulations. They include DAP discovery latency and effects of DAP failures to packet delivery rate and recovery latency. The results in [31] have demonstrated that the proposed solution exhibits self-organizing properties and therefore is appealing from a deployment perspective.

In [32], a simulation-based performance evaluation of RPL in real-life topology with empirical link quality data is presented. This study focuses on the mechanisms that RPL employs to repair link or node failures. Global repair is implemented by the DAG root with the help of periodic transmission of new DAG sequence number. As for local repair, a node will try to quickly and locally find an alternate parent upon the loss of the original parent. Results in [32] show that the network fixes local connection outage much quicker when local repair is jointly used with global repair then when only global repair is used. However, there are a few incidents, mainly in cases where packet delivery ratio is low or when DIS or DIO is not heard for a long time, where the outage time becomes comparable to the DAG sequence number period. The behavior and performance of these two mechanisms thus need further study and improvement for outdoor and large-scale networks like NANs.

The operation and performance of GPSR and RPL in NAN scenarios are compared in [33]. Extensive simulations are carried out to identify the advantages and disadvantages of each algorithm. Preliminary results demonstrate that RPL performs better than GPSR, however, the former requires a signaling mechanism and extra overhead for link quality estimation and a higher computation complexity for graph maintenance and path determination.

It is observed that even though RPL possesses many advanced features (objective function to capture channel dynamics, self-organizing coordinate to route traffic and so on), it requires heavy signaling to gauge link/network conditions and to propagate information that is necessary to construct and maintain the routing tree properly. Unless there is a mechanism that can efficiently control the signaling message broadcast procedure, significant overheads required by RPL could degrade the network performance. Routing fluctuation due to frequent changes in estimated link/path quality could be another issue with RPL. GEO, on the other hand, is very simple and truly distributed. It fully exploits the location information that is naturally available in NANs. Unfortunately, this kind of routing protocol has not been extensively studied in literature except for the work in [24–26]. Preliminary results are presented in [24] but there are three important limitations. First, an over-simplified free-space propagation channel model is assumed while real-life NANs are always deployed in a challenging outdoor environment with many factors that complicate the radio signal transmission. As a result, this assumption hinders the usefulness of results presented in [24]. Second, only message transmission reliability is measured. Transmission latency, which is one of the decisive performance metrics to be investigated in smart metering scenarios, is unfortunately not taken into consideration. Third, only conventional smart metering data is assumed in [24]. In fact, many advanced SG applications (e.g., distribution automation, fault detection and restoration and so on) are emerging. The work in [25,26] complement the work in [24] by investigating the performance of geographic routing protocols in practical NAN scenarios. GPSR and RPL are compared in [33]. Unfortunately, the resilience of these two protocols in the presence of network element failures has not been investigated.

4.3   GPSR Protocol for NAN

4.3.1   Detailed Description

GPSR is one of the GEO protocols that use location-based hop-by-hop forwarding principle. Basically, in each hop, they attempt to forward the messages closer to the destination by considering the current the positions of the current node, its neighbors and the final destination. GF is the simplest form of GEO and there are several greedy routing strategies, as illustrated in Fig. 4.2. The prevalent GF method employs the notion of distance progress and it forwards the message to the neighbor that has the shortest geographical distance to the destination (node N_1 in Fig. 4.2). For the most forward within radius (MFR), progress is defined as the distance between the current node X and the projection of a neighbor onto the line connecting node X and the final destination D. The larger this distance is, the more progress the corresponding neighbor can make. MFR forwards the message to the neighbor that maximizes the progress towards D, i.e., node N_2 in Fig. 4.2. Compass routing selects the neighbor that has the minimum angle between itself and the destination, i.e., node N_3 in Fig. 4.2. There is another strategy called nearest with forward progress (NFP) which forwards the packet to the nearest neighbor that is closer to the destination (node N_4 in Fig. 4.2).

It is noted that greedy routing does not always guarantee message delivery even if there is a path from source to destination. This occurs when there is no one-hop neighbor closer to the destination than the forwarding node itself. This is called a local minimum. An example of this problem is illustrated in Fig. 4.3. In this example, node X does not have any neighbor within its transmission range that is closer to destination D than X itself. As a result GF fails to find the next hop to forward the message even though there exists a possible path from X to D.

GPSR is one of the proposed solutions to handle the local minimum problem. It uses GF to forward packets to nodes that are always progressively closer to the destination. In regions of the network where such a greedy path does not exist,

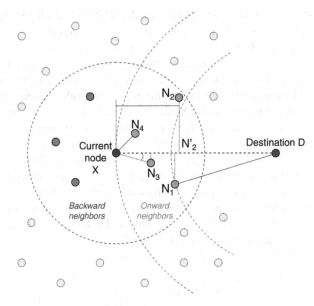

Fig. 4.2 Examples of different GF strategies: (N_1) = shortest geographical distance, (N_2) = MFR, (N_3) = compass routing and (N_4) = NFP

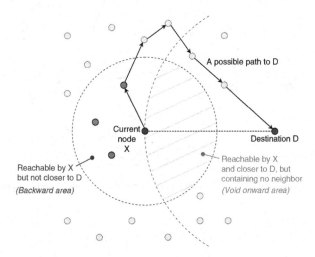

Fig. 4.3 An example of void area: there does not exist any node (within the transmission range of node X) closer to destination D

GPSR recovers by forwarding in perimeter mode. At the point where a node closer to the destination is found, GPSR switches back to greedy mode.

In perimeter mode, GPSR performs a simple planar graph traversal by employing the right hand rule (or the left hand rule). Suppose the mode changes to perimeter at node X for a message destined to destination D (see Fig. 4.4). From here on, the message is forwarded by employing the right hand rule, traversing the faces intersecting line XD which connects X and D. On each face, the traversal continues until the message reaches an edge that crosses line XD. At that edge, the packet moves to an adjacent edge, the first of which is determined by simply choosing the edge lying in counterclockwise direction from the intersected edge. Thereafter, as mentioned, the packet is forwarded around that face using the right hand rule. Figure 4.4 shows an example of perimeter forwarding starting at node X and ending at node Y where greedy mode is resumed. More detailed descriptions of GPSR can be found in [17]. Existing work shows that, compared to the conventional GF, GPSR can improve the packet delivery ratio [2].

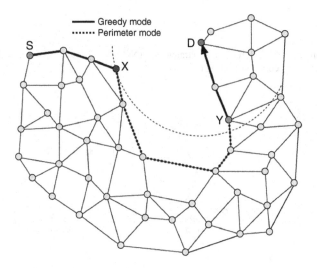

Fig. 4.4 An example of GPSR: the combination of greedy and perimeter modes to circumnavigate the void area

4.3.2 Advantages and Disadvantages

The advantage of GEO protocols is that it can achieve network wide routing while maintaining only neighborhood information at each node. The simplicity of GEO leads to good scalability since there is no need to keep routing tables up-to-date and to have a global view of the network topology and its changes. GEO protocols allow routers to be nearly stateless because forwarding decisions are based on location information of the destination and the location information of one-hop neighbors. No routing table needs to be constructed or maintained. A new node can join the network easily by locally exchanging information with existing nodes in its vicinity. Since establishment and maintenance of routes are not required, signaling overhead and computational complexity of GEO can be kept at a considerably low level. In addition to these advantages, the fact that locations of NAN devices are fixed and accurately known promotes GEO protocols as one of promising solutions for NANs.

However, since geographic distance cannot capture wireless channel conditions, the major flaw of GEO protocols is that the selection of next hop merely based on geographic distance may lead to void areas. These voids, local minima where there are no neighbors available that are closer to the destination, may inhibit forward progress of packets resulting in the failure of the forwarding strategy. Even though some variants of GEO protocols attempt to resolve this issue by having alternative forwarding mechanisms to go around the voids, e.g., the perimeter mode adopted in GPSR, lengthened routing paths may waste channel resource and thus degrade system performance.

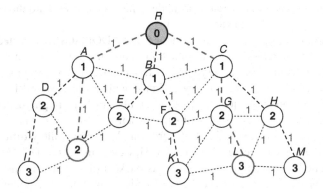

Fig. 4.5 An example DODAG using the hop count routing metric

4.4 RPL Protocol for NAN

4.4.1 Detailed Description

RPL belongs to the self-organizing coordinate routing class that constructs a viable coordinate system based on communication distance rather than the geographic distance used in location-based routing. RPL's key concept is the destination oriented directed acyclic graph (DODAG), a tree structure that specifies the routing paths between the root and the remaining nodes. The root is typically a gateway that acts as a common transit point that bridges every node and a backbone network [34].

Each node in the DODAG is assigned a rank that represents the cost of reaching the root as per the objective function (OF). The OF is designed to guide traffic to the root over paths that minimize a particular routing metric, such as hop count or expected transmission count (ETX). A list of possible metrics that could be used for the OF in RPL is presented in [35]. The rank of a given node is calculated based on the ranks of its neighbors, the cost to reach each of these neighbors and other routing metrics. Initially, the DODAG root starts sending out DAG information option (DIO) messages with a predefined lowest rank indicating that it is the traffic sink. Upon receiving a DIO, each node calculates its own rank based on information carried in the message and its local state. Each DIO contains information about the DODAG identification, the rank of the broadcasting node, parameters specifying the OF and so on. DIO's are periodically broadcasted from each node, triggered by the trickle timer. In this way, DIO's are gradually propagated from the root down to the most distant nodes and thereby help create a DODAG representing the physical network. For a given node, any neighbor with a lower rank is considered as a parent. When a node receives a packet destined to the root, it forwards the packet to its preferred parent. This results in the most cost-effective path to the root. In case no parent is available, the node can forward the packet to a sibling (equally ranked as the node itself).

For illustrative purposes, Figs. 4.5 and 4.6 show DODAGs constructed over the sample physical wireless network with hop count and ETX as routing metrics, respectively. In the first case, every wireless link connecting two neighboring nodes (denoted by a dashed line) is simply assigned with equal cost of "1" (hop) and the rank of each node is equal to the total number of hops required to reach the root starting from the node itself. Paths from each node to the root are denoted by thick dashed lines. For example, traffic from node J may follow different paths to root node R as can be seen in Fig. 4.5. However, using the DODAG and the aforementioned forwarding rule, J sends its packets to A which in turn forwards to R. This is the lowest-cost (i.e., 2-hop) path from A to R. Another example is the 3-hop path from L: $L \rightarrow G \rightarrow C \rightarrow R$. In the second case, hop count is

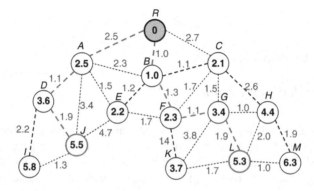

Fig. 4.6 An example of DODAG using the ETX routing metric

replaced by ETX, a measure of the quality of the wireless link between two nodes. Be definition, ETX is the number of expected packet transmissions necessary for error-less reception at the intended receiver. An ETX of one indicates a perfect transmission medium, whereas an ETX of infinity represents a completely non-functional link. For example, if it took n transmissions to successfully deliver m packets, the ETX of the link is $\frac{n}{m}$ ($n \geq m$). Due to varying characteristics of the transmission medium, ETX may vary widely from one link to another. As illustrated in Fig. 4.6, selected paths for traffic originating from J and L are $J \rightarrow D \rightarrow A \rightarrow R$ (5.5 transmissions) and $L \rightarrow G \rightarrow F \rightarrow B \rightarrow R$ (5.3 transmissions), respectively. By using ETX to reflect radio communication cost rather than simple hop count, this DODAG is expected to result in better system performance.

In order to control and limit signaling overhead, the duration of DIO broadcasts is doubled after each trickle timer expiration. The smallest interval between two consecutive DIOs is denoted by I_{min}, and number of times I_{min} can be doubled before maintaining a constant rate is denoted by $I_{doubling}$, so $I_{max} = I_{min} \times 2^{I_{doubling}}$. On any event that causes a change in the DODAG structure (parent node unreachable, new parent selection, new DAG sequence number, etc.), the timer is reset to I_{min}.

4.4.2 Advantages and Disadvantages

RPL exhibits many advantages that are desirable in the NAN setting. First, the tree-like graph constructed by RPL reflects the physical deployment and communication model of SMs (nodes) and DAPs (sinks or roots). Next, RPL is designed with the ability to incorporate various types of routing metrics and constraints that can be addictive, multiplicative, inclusive, exclusive and so on. Therefore, both QoS-aware and constraint-based routing disciplines can be supported. For instance, the ETX metric could be used to find the path that minimizes the total number of transmissions from nodes to the DAP while a constraint could be used to eliminate nodes that are battery-powered. Third, RPL allows multiple DODAG's to operate concurrently and independently to provide QoS differentiation for different classes of traffic in the NAN.

Primary requirements for the proper operation of RPL are the availability of network state information and control message propagation. This means that protocols to capture real-time status of wireless links, nodes and so on need to be designed and incorporated within RPL. Those protocols will obviously introduce signaling overhead in addition to overheads associated with control messages (DIS, DIO, etc.) propagation. Additionally, the time-varying nature of network state information may cause fluctuations in the structure of DODAGs. These fluctuations may then result in unnecessary changes in routing paths or even routing loops.

4.5 Proactive Parent Switching for RPL

Robustness is one of the key requirements when developing routing protocols for NANs. As the network topology changes due to network element (meters or wireless links) failures, it is imperative to dynamically update the routing decision. The reaction should be sufficiently fast to capture these changes. However, over-reacting could potentially compromise routing stability. Even though the robustness of RPL is addressed and studied in [32,34,36], the employed global and local repair mechanisms are quite simple and preliminary. Global repair is simply driven by a timer and it is mainly used to refresh the entire DAG and remove inconsistencies or loops that may appear over long operating periods (minutes or hours) rather than dealing with small-scale variations. The local repair is triggered only after a node loses its parents and siblings. It can help to fix local issues introduced by node failures or link fluctuations, however, due to the fact that the repair operates at the network layer (with the involvements of control message exchanges, node rank re-computations, parent-child relationship reforming, etc.), there could be significant delay and many packets may be dropped during the outage period. Therefore, this work proposes the PPS mechanism, which can effectively help RPL deflect network traffic from points of failures before local repair is activated. Instead of waiting until a node either detects the loss of its preferred parent (after exhausting all transmission/re-transmission attempts allowed by the MAC layer protocol) or runs out of all of its parents and siblings, PPS proactively nominates another parent as the next hop once the preferred parent is unreachable. Neighbor information supplied by the network layer is exploited to support the next-hop switching procedure performed at the MAC layer. Extensive simulations are carried out to demonstrate the effectiveness of PPS.

4.5.1 A Review on Global and Local Repair

In addition to the trickle timer and DIO messages to support DODAG formation, local and global graph repair mechanisms are utilized with the aim of repairing the network topology in case of link fluctuations or node failures. Local repair only causes partial DODAG changes while global repair affects all DODAG nodes [34].

The DODAG root governs the global repair operation by periodically increment-ing the DODAG version number. This initiates DODAG reconstruction. Nodes in the new DODAG version can choose a new position whose rank is not constrained by their rank within the old DODAG version. Global repair attempts to eliminate node rank inconsistencies, loops, and floating sub-graphs that may be present in the DODAG after long operating periods in order to re-optimize the entire routing hierarchy.

Local repair can be activated by any node that detects link fluctuations or node failures. It aims to find an alternate local path instead of globally re-optimizing the entire DODAG. RPL has two prominent local repair methods. The first one allows routing through alternate parents or siblings once the preferred parent is found to be lost. The second method, known as "poisoning", is used when a node runs out of parents and siblings. The node sends out a "poison" message to its children in order to detach itself from them. Then, it broadcasts a DODAG information solicitation (DIS) message soliciting DIO messages from all surrounding nodes thereby searching for nodes that can serve as its parents.

4.5.2 Proactive Parent Switching (PPS)

Neighbor information supplied by the network layer is exploited to support the next-hop switching procedure performed at the MAC layer. The operation principles of PPS imply that the reaction is triggered quickly so as to mitigate delay and waste of channel capacity due to useless back-off stages and re-transmissions during the outage period. Meanwhile, over-reaction to transient fluctuations of network elements is avoided by observing over a window of multiple MAC-layer transmission attempts. It is noted that the "poisoning" local repair and global repair specified in [32, 34, 36] can be used jointly with PPS for local DODAG re-construction and global DODAG re-optimization, respectively.

The operation of PPS is described as follows. At node i, upon receiving a data unit from the network layer, the MAC protocol first attempts to deliver the respective MAC protocol data unit (MPDU) to the preferred parent $P_{i,1}$. If this MPDU cannot be successfully delivered to $P_{i,1}$ after k_1 transmissions, an alternate parent, denoted by $P_{i,2}$, is attempted with k_2 transmissions. This procedure is iterated with maximum n parents (including the preferred parent) of node i. If all attempts fail, a failure notification will be sent to the network layer and the packet will be dropped. The maximum total number of transmissions that n parents attempt, denoted as K is,

$$\sum_{j=1}^{n} k_j = K. \tag{4.1}$$

Since, according to the routing rules specified by RPL, forwarding packets to a parent with a lower relative rank tends to result in a routing path that can reach the root at a lower cost, PPS should give the highest priority to the preferred parent $P_{i,1}$, then the first alternate parent $P_{i,2}$, then the second alternate parent $P_{i,3}$ and so forth. Note that the relative rank of parent $P_{i,j}$ is the sum of the rank of $P_{i,j}$ and ETX of the link from sender i to $P_{i,j}$. In order to enforce this priority, this work proposes that

$$k_u \geq k_v, \ \forall u < v; \ u, v \in \{1, 2, \ldots, n\}. \tag{4.2}$$

Algorithm 1 Proactive parent switching (PPS)

Require: Data unit D from the network layer, $\mathbf{P}_i$, $\mathbf{K}$
  Encapsulate D into MPDU
  for $(j \leftarrow 1; \; j \le n; \; j \leftarrow j + 1)$ **do**
    Update MPDU: MPDU.receiver $\leftarrow P_{i,j}, \ldots$
    for $(k \leftarrow 1; \; k \le k_j; \; k \leftarrow k + 1)$ **do**
      Transmit MPDU to $P_{i,j}$
      while (ACK timer not expired) **do**
        if ACK for MPDU is received from $P_{i,j}$ **then**
          Send confirmation to the network layer
          return [successful]
      Perform back-off procedure
  Send failure notification to the network layer
  return [failed]

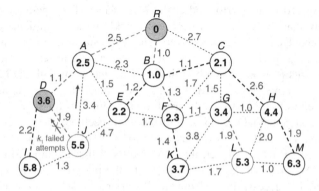

Fig. 4.7 RPL with PPS in a single node failure scenario

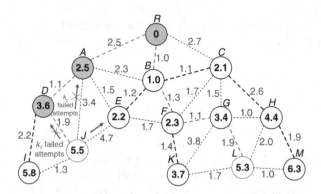

Fig. 4.8 RPL with PPS in a dual node failure scenario

The pseudo-code of PPS is presented in Algorithm 1 where $\mathbf{P}_i$ and $\mathbf{K}$ are the set of parents of node i and the set of maximum attempts for each corresponding parent, respectively, i.e., $\mathbf{P}_i = \{P_{i,1}, P_{i,2}, \ldots, P_{i,n}\}$ and $\mathbf{K} = \{k_1, k_2, \ldots, k_n\}$. It is noted that, for simplicity, not all operations carried out by the IEEE 802.11 MAC protocol are presented in Algorithm 1. The examples of RPL in cooperation with PPS are represented in Figs. 4.7 and 4.8. A single node failure scenario is illustrated in Fig. 4.7. Based on the constructed DODAG, node J intends to send data packets to its preferred parent, node D. However, since node D has failed, transmissions will not be acknowledged. Once the limitation of k_1 transmissions have been reached, node J has to switch its next hop to an alternate parent. If node J selects node A, its ETX value calculated from node A is $2.5 + 3.4 = 5.9$ transmissions. Whereas if node J selects node E, its ETX value calculated from node E is $2.2 + 4.7 = 6.9$ transmissions. Therefore, given that a lower ETX value for node J is desired, node A is selected to provide an alternative path , that is $J \rightarrow A \rightarrow R$. If node A has also failed, as shown in Fig. 4.8, after trying k_2 transmissions, node E becomes node J's last candidate parent and the network can recover from the failures of node D and A. The updated alternative path is $J \rightarrow E \rightarrow B \rightarrow R$.

4.5.3 Parameter Selection for PPS

The operation of PPS is governed by the selection of set **K**. In this study, $K = 7$ (commonly used for short IEEE 802.11 MAC frames) and $n = 3$ are assumed. To meet the constraint specified by (4.2), the simple well-known binary exponential rule is applied, i.e.,

$$\frac{k_j}{k_{j+1}} = 2, \ \forall j \in \{1, 2, \ldots, n-1\}. \tag{4.3}$$

Then, from Eqs. (4.1) and (4.3), the maximum numbers of transmissions for each of the three candidate parents are $\mathbf{K} = \{k_1, k_2, k_3\} = \{4, 2, 1\}$. This selected value of **K** will be used in the simulations presented in next chapters.

References

1. E. Alotaibi and B. Mukherjee, "A survey on routing algorithms for wireless ad-hoc and mesh networks," *Computer Networks*, vol. 56, no. 2, Feb. 2012.
2. T. Watteyne *et al.*, "From MANET to IETF ROLL standardization: A paradigm shift in WSN routing protocols," *IEEE Communications Surveys & Tutorials*, vol. 13, no. 4, pp. 688–707, Jan. 2011.
3. "Overview of existing routing protocols for low power and lossy networks," IETF ROLL, IETF draft, draft-ietf-roll-protocols-survey-07 (work in progress), Apr. 2009.
4. K. Akkaya and M. Younis, "A survey on routing protocols for wireless sensor networks," *Ad Hoc Networks*, vol. 3, no. 3, pp. 325Ǔ–349, Nov. 2005.
5. M. Abolhasan, T. A. Wysocki, and E. Dutkiewicz, "A review of routing protocols for mobile ad hoc networks," *Ad Hoc Networks*, vol. 2, no. 1, pp. 1–22, Jan. 2004.
6. J. N. Al-Karaki and A. E. Kamal, "Routing techniques in wireless sensor networks: A survey," *IEEE Wireless Commun.*, vol. 11, no. 6, pp. 6–28, Dec. 2004.

7. Y. C. Tseng, Y. S. C. S.-Y. Ni, and S. Jang-Ping, "The broadcast storm problem in a mobile ad hoc network," *Wireless Networks*, vol. 8, no. 2/3, pp. 153–167, 2002.

8. W. B. Heinzelman, A. P. Chandrakasan, and H. Balakrishnan, "An application-specific protocol architecture for wireless microsensor networks," *IEEE Trans. Wireless Commun.*, vol. 1, no. 4.

9. Z. Zhou *et al.*, "Energy-efficient cooperative communication in a clustered wireless sensor network," *IEEE Trans. Veh. Technol.*, vol. 57, no. 6.

10. O. Younis, M. Krunz, and S. Ramasubramanian, "Node clustering in wireless sensor networks: Recent developments and deployment challenges," *IEEE Network*, vol. 20, no. 3.

11. O. Younis and S. Fahmy, "HEED: a hybrid, energy-efficient, distributed clustering approach for ad hoc sensor networks," *IEEE Trans. Mobile Comput.*, vol. 3, no. 4.

12. A. Manjeshwar and D. P. Agrawal, "TEEN: a routing protocol for enhanced efficiency in wireless sensor networks," in *Proc. Workshop on Parallel and Distributed Computing Issues in Wireless Networks and Mobile Computing*, Apr. 2001, pp. 2009–2015.

13. ——, "APTEEN: A hybrid protocol for efficient routing and comprehensive information retrieval in wireless sensor networks source," in *Proc. International on Parallel and Distributed Processing Symposium (IPDPS)*, Fort Lauderdale, FL, USA, Apr. 2002, pp. 195–202.

14. A. Boukerche, R. W. Pazzi, and R. Araujo, "Fault-tolerant wireless sensor network routing protocols for the supervision of context-aware physical environments," *Journal of Parallel and Distributed Computing*, vol. 66, no. 4.

15. I. Stojmenovic and S. Olariu, *Handbook of Sensor Networks: Algorithms and Architectures*. Hoboken, New Jersey: John Wiley & Sons Inc., 2005.

16. P. Bose *et al.*, "Routing with guaranteed delivery in ad hoc wireless networks," in *Proc. 3rd ACM Int. Workshop on Discrete Algorithms and Methods for Mobile Computing and Communications (DIAL)*, Seattle, WA, USA, Aug. 1999, pp. 48–55.

17. B. Karp and H. Kung, "GPSR: greedy perimeter stateless routing for wireless networks," in *Proc. Annual International Conference on Mobile Computing and Networking (MobiCom)*, Boston, MA, USA, Aug. 2000, pp. 243–254.

18. H. Frey and I. Stojmenovic, "On delivery guarantees of face and combined greedy-face routing algorithms in ad hoc and sensor networks," in *Proc. 12th ACM Annual International Conference on Mobile Computing and Networking (MobiCom)*, Los Angeles, CA, USA, 2006, pp. 390–401.

19. E. Elhafsi, N. Mitton, and D. Simplot-Ryl, "End-to-end energy efficient geographic path discovery with guaranteed delivery in ad hoc and sensor networks," in *Proc. 19th Annual International Symposium on Personal, Indoor and Mobile Radio Communications (PIMRC)*, Cannes, France, 2008, pp. 1–5.

20. H. Kalosha *et al.*, "Select-and-protest-based beaconless georouting with guaranteed delivery in wireless sensor networks," in *Proc. 27th Conference on Computer Communications (INFOCOM)*, Phoenix, AZ, USA, Apr. 2008, pp. 346–350.

21. Q. D. Ho, Y. Gao, and T. Le-Ngoc, "Challenges and research opportunities in wireless communications networks for smart grid," *IEEE Wireless Communications*, pp. 89–95, Jun. 2013.

22. Q. D. Ho and T. Le-Ngoc. *Smart grid communications networks: Wireless technologies, protocols, issues and standards*, Chapter 5 in Handbook on Green Information and Communication Systems (Editors: S. O. Mohammad, A. Alagan, and W. Isaac), Elsevier, Summer 2012.

23. N. Saputro, K. Akkaya, and S. Uludag, "A survey of routing protocols for smart grid communications," *Computer Networks*, vol. 56, no. 11, pp. 2742–2771, 2012.

24. B. Lichtensteiger *et al.*, "RF mesh system for smart metering: System architecture and performance," in *Proc. IEEE Smart Grid Communication*, Mayland, USA, Oct. 2010, pp. 379–384.

25. G. Rajalingham, Q. D. Ho, and T. Le-Ngoc, "Evaluation of an efficient smart grid communication system at the neighbor area level," in *Proc. the 11th Annual IEEE Consumer Communications & Networking Conference (CCNC 2014)*, Las Vegas, USA, 9–13 Jan. 2014.

26. G. Rajalingham, Q. D. Ho, and T. Le-Ngoc, "Attainable throughput, delay and scalability for geographic routing on smart grid neighbor area networks," in *Proc. the 2013 IEEE Wireless Communications and Networking Conference (WCNC 2013)*, Shanghai, China, 7–10 Apr. 2013.

27. J. Ko *et al.*, "Evaluating the performance of RPL and 6LoWPAN in TinyOS," in *Proc. of the Workshop on Extending the Internet to Low power and Lossy Networks (IP+SN)*, Chicago, IL, USA, 2011.

28. O. Gnawali *et al.*, "Collection tree protocol," in *Proc. the 7th ACM Conference on Embedded Networked Sensor Systems (SenSys)*, Nov. 2009, pp. 1–14.

29. M. Nucolone, "Stability analysis of the delays of the routing protocol over low power and lossy networks," Master's thesis, KTH Electrical Engineering, 2010.

30. D. Wang *et al.*, "RPL based routing for advanced metering infrastructure in smart grid," Mitsubishi Electric Research Laboratories, Tech. Rep. TR2010-053, Jul. 2010.

31. P. Kulkarni *et al.*, "A self-organising mesh networking solution based on enhanced RPL for smart metering communications," in *Proc. IEEE International Symposium on a World of Wireless, Mobile and Multimedia Networks (WoWMoM)*, Jun. 2011, pp. 1–6.

32. J. Tripathi, J. C. de Oliveira, and J. P. Vasseur, "Applicability study of RPL with local repair in smart grid substation networks," in *n Proc. the First IEEE International Conference on Smart Grid Communications (SmartGridComm)*, Oct. 2010, pp. 262–267.

33. Q.D Ho, Y. Gao, G. Rajalingham, T. Le-Ngoc, "Performance and applicability of candidate routing protocols for smart grid's wireless mesh neighbor area networks," in *Proc. the 2014 IEEE International Conference on Communications (ICC 2014)*, Sydney, Australia, 10–14 Jun. 2014.

34. "RPL: IPv6 routing protocol for low-power and lossy networks," Internet Engineering Task Force (IETF), RFC 6550, March 2012.

35. "Routing metrics used for path calculation in low-power and lossy networks," Internet Engineering Task Force (IETF), RFC 6551, March 2012.

36. K. D. Korte, A. Sehgal, and J. Schonwalder, "A study of the RPL repair process using ContikiRPL," in *Proc. the 7th International Conference on Autonomous Infrastucture, Management and Security (AIMS)*, Jun. 2012, pp. 50–61.

Chapter 5
Performance and Feasibility of GPSR and RPL in NANs

Given the two promising routing protocols for wireless mesh NANs, namely GPSR and RPL, this chapter presents extensive simulations carried out with IEEE 802.11-based radio and practical system parameters related to the NAN's characteristics and deployment scenarios specified in SG standards. For the robustness of routing protocols against network element failures, this chapter investigates the capability of PPS in rerouting traffic around non-connected network regions caused by SM malfunctions. The obtained results reveal the advantages of RPL against the GPSR in terms of its network condition awareness and superior performance. They also demonstrate that PPS can effectively improve the network resilience by adaptively forwarding the traffic over multiple alternative paths.

5.1 Simulation Setup

5.1.1 Network Topology and SM Placement

A discrete event network simulation platform is used to simulate NAN clusters each consisting of one DAP and n network nodes (i.e., SMs). All nodes are assumed to be homogeneous except for the DAP which has an additional interface to communicate with the upper tier network (e.g., LTE backhaul). In this section, both network node and SM are synonymous and refer to a digital power electric meter with communications capability and other smart features (e.g., data processing, computing, etc.). Each node has a radio communications module responsible for relaying sensor data (e.g., voltage, current, phase, energy consumption and so on) to the DAP (either directly or via other SMs using multi-hop paths). This means that each SM acts not only as a network end-device but also as a router to relay traffic for other SMs and form a mesh network. In order to ensure that the results obtained in this work are meaningful and applicable to real-life scenarios, parameters related

© The Author(s) 2014
Q.-D. Ho et al., *Wireless Communications Networks for the Smart Grid*,
SpringerBriefs in Computer Science, DOI 10.1007/978-3-319-10347-1_5

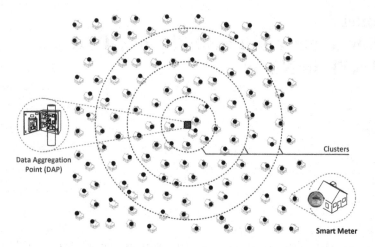

Fig. 5.1 Illustrative NAN clusters with varying sizes

to SM deployment scenarios and wireless channel characteristics specified by the National Institute of Standards and Technology's SG Priority Action Plan 2 [1] are taken into consideration. Specifically, typical SM density in urban areas is assumed, i.e., $\rho = 2{,}000$ SMs per square kilometer. Figure 5.1 visualizes simulated NAN clusters with different sizes each delimited by a dashed circle. Since SM density is unchanged, cluster size (or scale) is represented either by the number of SMs, n, located within the cluster or by the radius, R, of the circle which sets the limit for the cluster. The relation between these two parameters is represented by $n = \pi R^2 \rho$. The DAP is represented by a red square located at the center of the clusters while SMs are represented by black dots uniformly distributed in the area.

5.1.2   Wireless Channel Model

Following the specifications in [1], the wireless channel is modeled by path-loss (with path-loss exponent α) and log-normal shadowing X_σ (with standard deviation σ) as shown below:

$$P_R[\mathrm{dBm}] = P_T[\mathrm{dBm}] + 10\log_{10}\left[\frac{\lambda^2}{16\pi^2 d^\alpha}\right] - X_\sigma, \qquad (5.1)$$

where P_T and P_R are the transmitted and received radio power, respectively; λ and d are the wavelength and the transmitter-receiver distance, respectively. In simulations, transmitted radio power P_T is chosen to have the communication range of approximately 50 m. The value for path-loss exponent α is taken from [1] and all parameters related to wireless channel are listed in Table 5.1.

Table 5.1 Simulation model and parameters

Meter deployment	Node density	$\rho = 2{,}000$ nodes/km^2
	Node placement	Uniformly random
	Num. of nodes per cluster	n (varying)
Wireless channel	Path-loss	$\alpha = 3.6$
	Shadowing	Log-normal, σ (varying)
PHY layer	Standard	IEEE 802.11b
	Frequency band	2.4 GHz
	Transmission rates	$\{1.0, 2.0, 5.5, 11.0\}$ Mbps
MAC layer	Standard	IEEE 802.11b
	Operation mode	Mesh
	RTS/CTS	Disabled
	ACK	Enabled
	Max. retransmissions	7
	Back-off procedure	Binary exponential
	Min. contention window	$CW_{min} = 31$
	Max. contention window	$CW_{max} = 1{,}023$
	ST, SIFS, PIFS, DIFS, EIFS	20, 10, 30, 50, 364 μs
APP layer	Data length	$L_0 = 100$ bytes per packet
	Packet rate	r packets/s (varying)

PIFS point coordination function interframe spacing, *DIFS* distributed coordination function interframe spacing = 2 × ST + SIFS, *EIFS* extended interframe spacing = SIFS + DIFS + Total ACK length/PHY header rate

5.1.3 PHY and MAC Layer Specifications

For the communications modules built into each SM, IEEE 802.11b physical (PHY) and medium access control (MAC) layers are selected. Simulation parameters related to these two layers are summarized in Table 5.1. The PHY layer operates in the 2.4 GHz frequency band and uses direct sequence spread spectrum (DSSS) technology. Adaptive modulation and coding (AMC) can support multiple data transmission rates, i.e., 1.0, 2.0, 5.5, or 11.0 Mbps, depending on channel conditions. The MAC layer employs the carrier sense multiple access with collision avoidance (CSMA/CA) mechanism. A node with a new packet to transmit will continuously sense the channel in the hope that it remains idle for a time interval equal to a distributed coordination function interframe spacing (DIFS). When the channel is measured idle for a DIFS, the node backs off for a random period of time. After expiry of the back-off time, the node transmits if the channel is still idle. This back-off mechanism attempts to minimize the probability of transmission collision. In addition, to avoid channel capture, a node must wait a random back-off time between two consecutive new packet transmissions, even if the medium is sensed idle in the DIFS time. For each packet transmission, the back-off time is X times the contention window slot time (ST) where X is picked uniformly in

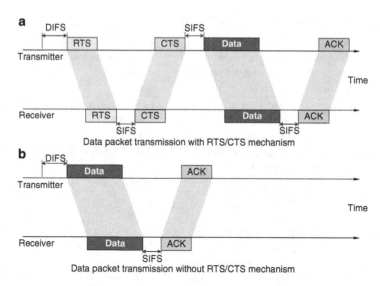

Fig. 5.2 The timing diagram showing basic operations of IEEE 802.11 CSMA/CA with and without RTS/CTS

$\{0, 1, \ldots, C W_n\}$ where $C W_n$ represents the contention window which is a function of the number of failed packet transmissions. At the first transmission attempt, $C W_0$ is set to equal the minimum contention window $C W_{min}$. Binary exponential back-off is assumed: after each unsuccessful transmission, the contention window is doubled, up to a maximum value $C W_{max}$. In other words, the contention window for the n-th trial is $C W_n = \min\{(C W_{min} + 1) \times 2^n - 1, C W_{max}\}$. The back-off time counter is decremented and a node only transmits when the back-off time reaches zero. The receiver sends an ACK signal, after a period of time called the short interframe spacing (SIFS), to signal successful packet reception. If the transmitting node does not receive the ACK, it reschedules packet transmission according to the given back-off rules. Request-to-send (RTS)/Clear-to-send(CTS) mechanism is optional. A timing diagram illustrating the transmission of a data packet using IEEE 802.11 CSMA/CA with RTS/CTS is given in Fig. 5.2. More details on IEEE 802.11 PHY and MAC layers can be found in [2].

5.1.4 Traffic Model

Since SMs are deployed to support not only conventional SG applications (e.g., meter reading, power outage detection, demand response and so on) but also emerging future applications (e.g., advanced distribution automation, fault detection and restoration, electric vehicles and so forth), they are expected to handle the exchange of an increasing volume of information. As a result, in this work, application (APP) data traffic generated by each node in the simulated NAN cluster is swept over a wide range to gauge how the performance of the cluster scales under different levels of offered load. APP data packets are generated periodically at rate r [packets/sec/node]. The APP data length L_0 is assumed to be 100 bytes. It is noted that, in this work, only uplink traffic (i.e., from nodes to the DAP) is considered since the communications in this direction is converge-cast in nature and more challenging in the NAN scenario, as compared to that of the downlink direction. For conventional SG applications including interval/on-demand meter reading, demand response and remote connect/disconnect applications as mentioned in NIST's PAP02 [1], following the estimation given in Table 7 [1], the total traffic load offered to the network is approximately 16,808 bytes/m/day. Since it is assumed in this work that each packet carries 100 data bytes, the equivalent data packet rate is $r = \frac{16,808}{24 \times 60 \times 60 \times 100} \approx 0.00195$ packets/s/node, hereafter referred to as the base data rate.

5.2 Performance Metrics

The performance of the network is mainly analyzed based on data packet transmission delay and delivery ratio. Besides, in order to have an in-depth understanding of each algorithm, some other parameters are also investigated. They include routing path length (in terms of hop count) and mean transmission time (MTX).

5.2.1 Packet Transmission Delay

Transmission delay D_p of data packet p accounts for the duration from the time when p is ready for the transmission at the original source until p is received and decoded correctly at its final destination. D includes various components: the time for sending and receiving control messages (T_p^{RTS}, T_p^{CTS}, T_p^{ACK}) and packet p itself (T_p^{DATA}), back-off time (T_{BO}) and interframe spacings (T_{IFS}). In other words, its calculation is as follows:

$$D_p = T_p^{\text{RTS}} + T_p^{\text{CTS}} + T_p^{\text{ACK}} + T_p^{\text{DATA}} + T_p^{\text{BO}} + T_p^{\text{IFS}}. \tag{5.2}$$

Note that these components represent accumulated values since for one successful data packet delivery there might be multiple back-off stages and packet re-transmissions.

5.2.2 Packet Delivery Ratio (PDR)

PDR, denoted by P_D, is the percentage of data packets that are received and decoded successfully by the DAP. It is calculated by the following equation:

$$P_D = \frac{|\mathbf{N_{rx}}|}{|\mathbf{N_{tx}}|},\qquad(5.3)$$

where $\mathbf{N_{tx}}$ is the set of data packets generated and sent by network nodes. In each simulation setting, $|\mathbf{N_{tx}}|$ is chosen to be 100,000 for statistical measurement of P_D.

Note that statistics of delay and PDR can be obtained for the entire simulation or per node for more detailed investigations. In addition to these two primary performance metrics, hop counts of routing paths, number of neighbors for each node and so on could be analyzed to give an in-depth understanding of GPSR.

5.3 Simulation Results and Discussions

Existing studies show that the performance of WMNs varies with network scale, wireless channel characteristics, deployment settings, traffic loading levels and equipment availability. Therefore, in order to assess the feasibility of GPSR and RPL routing protocols in wireless mesh NANs, this section provides a comprehensive study on the performance of these two protocols in practical networks with parameters specified by SG communications standards. Specifically, two main scenarios are investigated.

The first scenario assumes that all network nodes are active and operate without failure during the entire simulation time. PDR and transmission delay are studied to capture the system performance of GPSR and RPL and its trend when the variance of channel shadowing σ, per-node data rate r and cluster size n are swept. It is expected that increasing channel randomness will results in more packet corruptions and thus lead to a higher rate of packet losses, re-transmissions and in turn performance degradation. When r increases, packets are generated and sent more often and thereby induce a higher chance for channel contentions, back-offs and packet re-transmissions. Even though routing paths are not lengthened, longer delays and lower transmission reliability levels should be observed. Cluster size n is an important parameter in system design since the larger it is, the lower the required costs related to installation, operation and maintenance of DAPs. However, an increase in the cluster size will have a dual effect. First, a large number of nodes in the cluster will inject more traffic towards the DAP. Second, since node density is fixed ($\rho = 2{,}000$ nodes/km^2) and $n = \pi R^2 \rho$, a larger number of nodes also implies that the geographic area of the cluster, i.e., πR^2, expands. Packets from nodes farther away need to traverse longer distances (or more hops) to reach the DAP. This will effectively increase the network load and average hop count at the same time. As a result, this dual effect should significantly increase transmission delay while decreasing PDR.

Table 5.2 Simulation parameters of study cases I, II, III and IV

Parameter	Study case			
	I	II	III	IV
Shadowing σ (dB)	0	0,4,8,12	8	8
Packet rate r (packet/s/node)	0.00195	0.00195	**0.001, 0.00195, 0.01, 0.1**	0.00195
Cluster size n (nodes)	1,000	1,000	1,000	**1,000, 3,000, 6,000**
Network availability a (%)	100	100	100	100

For the detailed implementation of GPSR, hello messages are sent at the same rate as data messages. The neighbor table is refreshed after every iteration of two hello message broadcasts. RPL is implemented based on the design principle of the original protocol along with the default parameters specified in [3] without the PPS mechanism. The DIO generation and transmission are controlled by a unique trickle timer at each node whose parameters are specified by [3] as well. If the routing topology is not consistent, such as a node changes its preferred routing path or a new node joins the DODAG, the trickle timer resolves the inconsistency by resetting its interval to the minimum value. This is done in order to generate more frequent DIOs and disseminate updated information within the DODAG. Simulation results presented in this section will investigate the operation and compare the performance of these two protocols over a wide range of system parameters. The study cases carried out in this section are briefly described as follows. Study case I is to verify the operations of GPSR and RPL under an ideal channel condition. Study case II, III and IV investigate the effects of channel condition, network offered load and cluster size on the system performance, respectively. The two protocols are contrasted and explained in each study case. Table 5.2 summarizes the four study cases.

In the second scenario, the robustness of GPSR and RPL is evaluated by taking into account the effect of node failures. The fraction of nodes that randomly fail during the simulation time is represented by $f \in [0, 1]$. Failed nodes and arrival time of their failures are uniformly distributed. A node remains silent, i.e., incapable of transmitting and receiving any packet, after it fails. Availability of the network is thus given by $a = 1 - f$. In order to address the impact of node failures, the robustness of GPSR, conventional RPL and RPL with the proposed PPS mechanism are compared in scenarios with different levels of network availability. The decrease of network availability a will result in an increasing number of network disconnections and communication truncations between nodes. Larger number of re-transmissions and control packets are imposed on the network to reduce packet loss by detecting and recovering from node failures. The related results and observations will be presented. Furthermore, with the aim of understanding the impact of growing traffic load in the occurrence of node failures, the per-node data rate is also swept. Two study cases will be considered as presented in Table 5.3.

Table 5.3 Simulation parameters of study cases V and VI

Parameter	Study case V	VI
Shadowing σ (dB)	8	8
Packet rate r (packet/s/node)	0.00195	**0.001, 0.00195, 0.01, 0.1**
Cluster size n (nodes)	1,000	1,000
Network availability a (%)	**100, 95, 90**	90

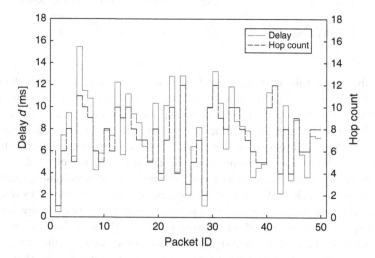

Fig. 5.3 Delay and hop count for 50 sampled packet transmissions (GPSR, case I)

Study case V compares the robustness of GPSR and RPL with PPS against the rate of node failures. Per-node data rate is swept in study case VI to investigate the effects of both node failure and traffic load.

5.3.1 Network Performance Without Node Failure

5.3.1.1 Study Case I: Protocol Operations Without Shadowing

In this study case, an ideal wireless channel, i.e., no shadowing, is assumed. GPSR and RPL are expected to have similar performance since the ETX routing metric used by RPL might not be able to provide any advantage when there is no channel randomness. The simulation parameters in this case are set as follows: channel shadowing $\sigma = 0.0\,\text{dB}$, $r = 0.00195$ packets/s/node (base data rate) and $n = 1,000$ nodes ($R = 398.9\,\text{m}$).

Simulation results and associated discussions will demonstrate and explain above-mentioned expectations as follows. First, Figs. 5.3 and 5.4 plot delays (represented by solid lines) and routing path lengths in terms of hop count

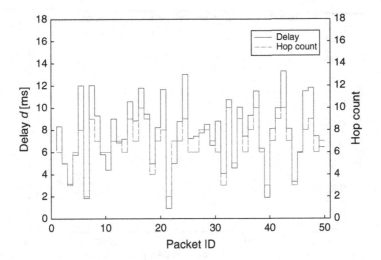

Fig. 5.4 Delay and hop count of 50 sampled packet transmissions (RPL, case I)

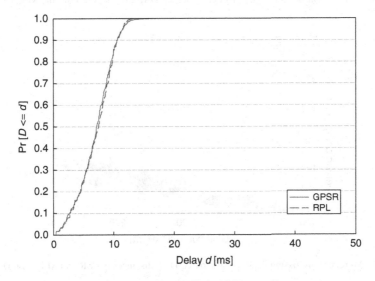

Fig. 5.5 CDF of packet transmission delays (GPSR and RPL, case I)

(represented by dashed lines) of 50 sampled packets received at the DAP for GPSR and RPL, respectively. It can be seen that packets routed over longer paths (traversing a large number of hops) to reach the DAP generally experience higher delay for both GPSR and RPL. However, the correlation between the hop count and the delay does not always hold since different transmissions may experience different channel conditions and thus different number of collisions, back-off stages and re-transmissions. The cumulative distribution functions (CDFs) of delays given by these two protocols are then given in Fig. 5.5. It can be seen that GPSR and RPL have quite similar packet delay statistics (the two curves are nearly identical).

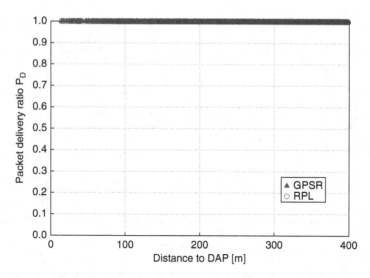

Fig. 5.6 Average transmission reliability versus source-DAP distances (GPSR and RPL, case I)

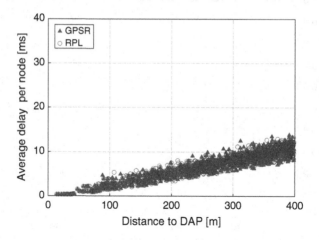

Fig. 5.7 Average transmission delay versus source-DAP distances (GPSR and RPL, case I)

For further investigations on the operations of these two algorithms, the correlation between source-destination distance and PDR, average transmission delay
and routing path length per node are plotted in Figs. 5.6–5.8, respectively. In these
figures, the horizontal axis represents the distance from source nodes to the DAP.
As can be seen in Fig. 5.6, every node can successfully deliver all of its packets to
the DAP. Even for nodes that are far away from the DAP (i.e., beyond roughly
400 m to the DAP), their P_D's are still 100 %. The average transmission delay
versus geographical distance from sources to the final destination is illustrated
in Fig. 5.7. For nodes that are located no more than 50 m away from the DAP,

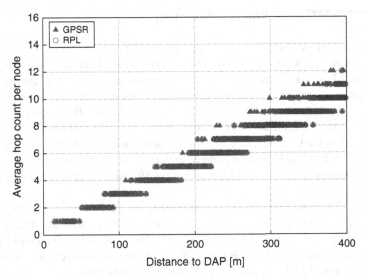

Fig. 5.8 Average hop count versus source-DAP distances (GPSR and RPL, case I)

almost all packets for GPSR and RPL have a delay of around 0.3526 ms. This can be explained by two facts: (i) these packets are transmitted directly to the DAP since source-destination distances are shorter than the 50 m transmission range; and (ii) no re-transmission is required since communication links over short distance are reliable. When source-destination distance is greater than the transmission range, multi-hop paths have to be employed to deliver packets to the DAP. Obviously, multi-hop transmissions result in longer delays. Moreover, most packets routed by GPSR and RPL experience quite similar average delays and they traverse the same number of hops to the DAP, as shown in Figs. 5.7 and 5.8. In this case where there is no channel shadowing, these results indicate that GPSR and RPL perform similarly, as expected.

The similarity of GPSR and RPL can also be explained as follows. When the link is ideal, packets are successfully delivered by greedy forwarding and GPSR approximates the shortest path routing with the minimum number of hops and the longest distances on each hop, as previously demonstrated in [4]. RPL is basically designed to minimize the total path cost (in terms of ETX) in order to ensure high packet reliability. It therefore may lead to relatively longer but more reliable routing paths than GPSR. However, in this case, wireless communications between each pair of nodes is ideal and always reliable within the communication range. Therefore, the ETX metric reflects the hop count. In other words, both RPL and GPSR tend to select similar paths that minimize the total hop count and thus have similar operation and performance.

5.3.1.2 Study Case II: Effects of Shadowing

In reality, wireless channels randomly change due to many factors such as shadowing and thus packet transmissions could be corrupted. Therefore, this study case investigates how GRPS and RPL perform when channel shadowing, quantified by the variance σ, varies. Specifically, the benefit of employing ETX routing metric in RPL is studied. First, a wireless channel with $\sigma = 8.0\,\text{dB}$ is considered as a representation of a realistic NAN, as specified in [1]. RPL, which selects routing paths based on the ETX metric, is expected to perform better than GPSR when there is channel shadowing. Parameters related to cluster and application data rate are the same as those of the study case I, i.e., $r = 0.00195$ packets/s/node and $n = 1,000$ nodes ($R = 398.9\,\text{m}$).

Figures 5.9 and 5.10 show that, compared to the case of no shadowing (study case I), system performance in terms of both transmission delay and reliability of GPSR and RPL is degraded. Further, RPL outperforms GPSR: (i) D_{95} given by RPL is much lower than that of GPSR (26.57 and 43.83 ms, respectively, as can be seen in Fig. 5.9) and (ii) RPL successfully delivers more than 99.82 % of packets to the DAP while GPSR only guarantees around 98.37 % (average transmission reliability per node as can be seen in Fig. 5.10).

More detailed comparisons between GPSR and RPL are presented in Figs. 5.10–5.12. As can be seen in Fig. 5.10, with GPSR, nodes farther away from the DAP generally suffer from lower reliability. Specifically, nodes that are beyond 200 m from the DAP suffer from noticeable packet loss. For example, for nodes that are located around 300 m from the DAP, only 92.45 % of their packets can reach the DAP. Meanwhile, with RPL, every node can still successfully deliver

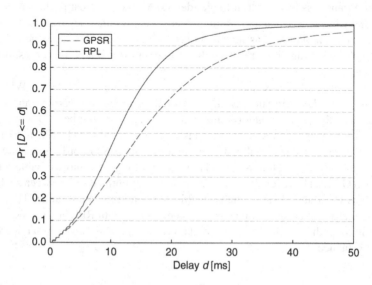

Fig. 5.9 CDF of packet transmission delays (GPSR and RPL, case II)

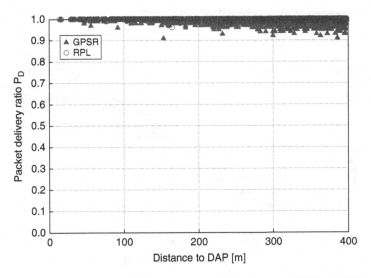

Fig. 5.10 Average transmission reliability versus source-DAP distances (GPSR and RPL, case II)

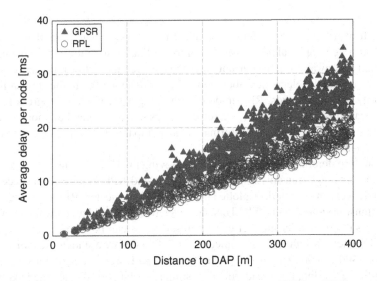

Fig. 5.11 Average transmission delay versus source-DAP distances (GPSR and RPL, case II)

nearly all of its packets to the DAP. Even for nodes that are quite far away from the DAP (i.e., beyond 300 m from the DAP), their P_D's are still higher than 96.50 %. Additionally, as shown in Fig. 5.11, average transmission delay per node given by RPL never exceeds 22.0 ms while it can be as high as 35.0 ms for GPSR. Figure 5.12 demonstrates that packets traverse over multi-hop routing paths. For nodes at the same distance to the DAP, their packets routed by RPL traverse a larger number of hops as compared to GPSR. This can be explained as follows. By using

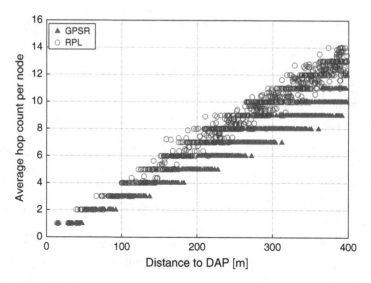

Fig. 5.12 Average hop count versus source-DAP distances (GPSR and RPL, case II)

greedy forwarding, GPSR attempts to select the next-hop neighbors that result in the greatest geographical progression towards the final destination. Those neighbors are usually far away from the current node and have relatively low link quality. RPL, on the other hand, selects neighbors with shorter distance but higher transmission reliability. In other words, RPL trade-offs the longer forwarding progress for the higher transmission reliability. Therefore, it results in longer but more reliable paths and thus better overall performance, as compared to GPSR, as observed in Figs. 5.10–5.12.

In order to demonstrate the benefit of employing the ETX routing metric to route traffic, the average transmission time that GPSR and RPL require for a successful packet delivery in each node is plotted in Fig. 5.13. As expected, RPL in fact requires fewer transmissions than GPSR. This, however, does not hold at a few nodes where GPSR results in smaller number of transmissions, as can be seen in the bottom of Fig. 5.13 where the curves are zoomed in for 200 nodes whose identifications range are from 400 to 600. This is due to the fact that RPL nominates preferred parents for traffic forwarding by considering the estimated total cost to reach the DAP from the current node (which is the sum of the link ETX from the node to the candidate parent and the rank of that parent), not merely on ETX of the link. This selection rule is applied to ensure that the traffic will finally reach the DAP with the lowest total cost.

Additionally, when selecting neighbors that are far away from the current node in order maximize the geographical progression towards the final destination, GPSR may unfortunately force the PHY layer to transmit the packets at low data rates and thus contribute to longer overall delays, as opposed to RPL. Figure 5.14 shows that GPSR only transmits 46.03 % of the packets at the highest PHY data rate

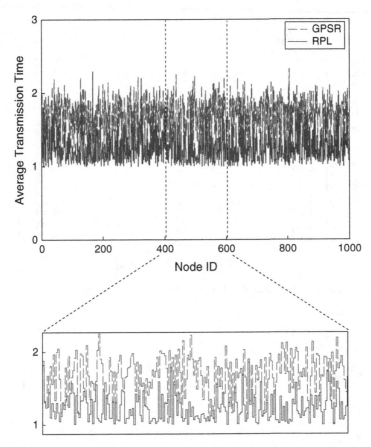

Fig. 5.13 Average transmission time per successful delivery for each node (GPSR and RPL, case II)

(11.0 Mbps) while that number for RPL is much higher, 64.46 %. In the other extreme, GPSR transmits 40.37 % of the packets at the lowest PHY data rate (1 Mbps) while that number for RPL is much lower, 25.1 %.

Next, in order to see how the system performance is degraded due to the increasing level of channel shadowing, P_D and D_{95} when shadowing variance σ is swept from 0.0 (no shadowing) to 12.0 dB (severe shadowing) are plotted in Figs. 5.15 and 5.16, respectively. When shadowing increases, the channel randomness increases accordingly and thus links are less reliable and the network performance degrades for both GPSR and RPL. For example, when σ doubles from 4.0 to 8.0 dB, for GPSR, the reliability is reduced from 99.10 to 98.37 % while D_{95} is increased from 34.5 to 43.83 ms. The significant increase in delay is the result of a higher rate of channel contention, back-off and re-transmissions. In this situation, acknowledgment and re-transmission mechanisms of IEEE 802.11 MAC layer help to maintain sufficiently high PDR. Similar effects are also observed with

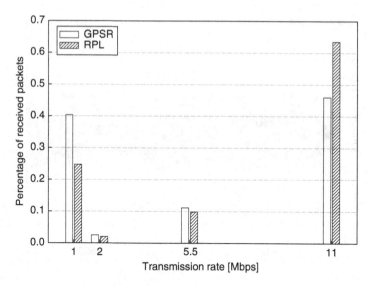

Fig. 5.14 Packet transmission rate distribution (GPSR and RPL, case II)

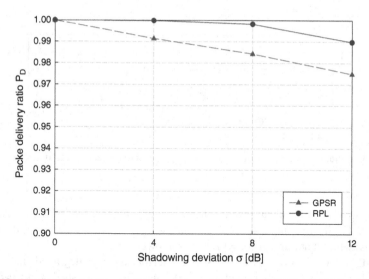

Fig. 5.15 Average packet delivery ratio versus channel shadowing (GPSR and RPL, case II)

RPL. When there is no shadowing ($\sigma = 0.0$ dB), as shown in study case I, GPSR and RPL perform similarly. For all shadowing cases (with positive σ), RPL outperforms GPSR in terms of both transmission delay and reliability.

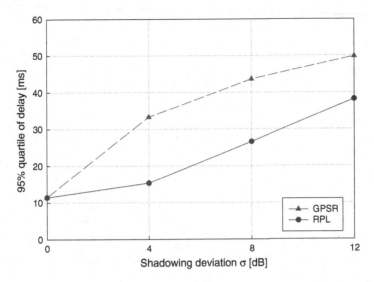

Fig. 5.16 Ninety-fifth percentile of transmission delay versus channel shadowing (GPSR and RPL, case II)

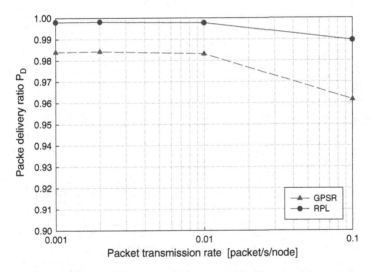

Fig. 5.17 Average packet delivery ratio versus per-meter data rates (GPSR and RPL, case III)

5.3.1.3   Study Case III: Effects of Data Traffic Load

Since SMs are deployed to support not only conventional SG applications (e.g., meter reading, demand response, and so on as mentioned in [1]) but also those expected in the future (e.g., advanced distribution automation, fault detection and restoration, etc.), they are responsible for exchanging an increasing volume of

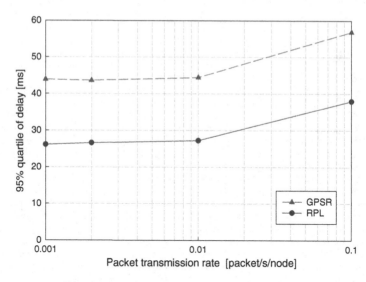

Fig. 5.18 Ninety-fifth percentile of transmission delay versus per-meter data rates (GPSR and RPL, case III)

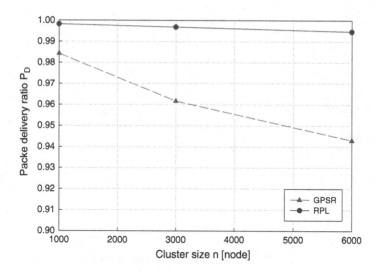

Fig. 5.19 Average packet delivery ratio versus cluster sizes (GPSR and RPL, case IV)

information. As a result, the simulation presented in this study case investigates how the network performance scales with network offered load. The system performance is studied when each node offers an increasing level of load to the network. Per-node data rate r is swept and shadowing variance σ and cluster size n are held constant at 8.0 dB and 1,000 nodes, respectively. When data packets are sent more often, channel contentions take place with a higher probability. This results in a large

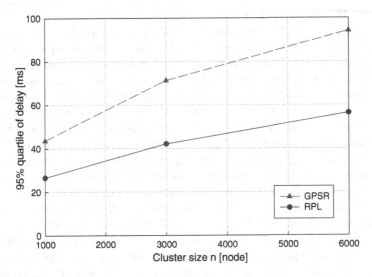

Fig. 5.20 Ninety-fifth percentile of transmission delay versus cluster sizes (GPSR and RPL, case IV)

number of back-off stages per packet. Even though routing paths are not lengthened (since cluster size is unchanged), lower transmission reliability and longer delays are observed for both GPSR and RPL. Figures 5.17 and 5.18 show that, as data rate increases, the performance of both GPSR and RPL degrades, fortunately, RPL still achieves lower transmission delay and higher PDR than GPSR. For example, it can be seen from Fig. 5.17 that when traffic rate r increases from 0.01 to 0.1 packets/s/node, P_D given by GPSR is reduced from 98.32 to 96.16 %, as compared to the decrease from 99.78 to 98.96 % of RPL. D_{95} of GPSR increases from 44.42 to 56.9 ms while that of RPL increases from 27.22 to 38 ms, as shown in Fig. 5.18. Furthermore, with heavier traffic loads, unreliable routing path selected by GPSR and its higher hello message frequency both lead to a significant increase in channel contentions thus more packet losses. Therefore, the gaps between RPL and GPSR in terms of PDR and transmission reliability widen when per-meter data rate r is increased from 0.001 to 0.1 packets/s/node.

5.3.1.4 Study Case IV: Effects of Cluster Size

Scalability is one of the most important issues of NAN communications. With the current trend for incremental SG deployment, the number of SMs will increase dramatically. Consequently, there is a great challenge for candidate routing protocols to provide an acceptable level of communications with a large number of nodes. This study case investigates the effects of cluster size on system performance of GPSR and RPL. Shadowing variance σ and per-node data rate r remain constant at 8.0 dB and 0.00195 packets/s/node. Figures 5.19 and 5.20 show the scalability of these two routing protocols. With the same node density, increasing the number of nodes leads to expanding the coverage area for each cluster. As the cluster size increases, network offered traffic load and average traffic path length are both increased. Since packets routed over a longer path are more likely to be lost, packet loss increases. When varying the number of nodes from 1,000 to 6,000, P_D of GPSR drops from 98.43 to 94.31 %, but RPL still maintains its P_D, i.e., from 99.82 to 99.47 %. The reliability benefits of employing RPL are enhanced in a larger network. The transmission delay is significantly higher with increasing network size for both RPL and GPSR. Packets with GPSR generally require fewer hops to reach the destination. However, long transmission hops experiences unreliable transmission and requires a larger number of re-transmissions for successful packet delivery. D_{95} of GPSR dramatically grows from 43.62 to 94.16 ms, whereas RPL still performs better with relatively lower D_{95}, e.g., 23.56–56.33 ms. RPL provides shorter delay than GPSR, even with longer traversed path in a larger network. This can be explained through the fact that a higher proportion of the transmission delay is due to exponential back-off delays as opposed to propagation delay due to multiple hops. Therefore, RPL is a more efficient and reliable routing protocol in large-scale networks.

5.3.2   Network Performance with Node Failures

This section focuses on investigating the effect of node failures on network performance and how PPS can help RPL mitigate them. The performance of RPL with $\mathbf{K} = \{4, 2, 1\}$, namely, RPL-PPS(4,2,1), is evaluated. In order to verify the parent switching principle proposed by PPS and demonstrate its advantages, GPSR and other three schemes of RPL as references are also studied: RPL-PPS(1,2,4) (i.e., $\mathbf{K} = \{1, 2, 4\}$: parents with higher relative ranks are given greater numbers of transmission attempts, as opposed to the proposed principle), RPL-PPS(1,1,...,1) (i.e., $\mathbf{K} = \{1, 1, 1, 1, 1, 1, 1\}$: the packet is switched to another parent once a single transmission attempt fails), and the conventional RPL (without using PPS), namely RPL-non-PPS.

5.3.2.1   Study Case V: Effects of Node Failures

Communications in NANs requires the routing protocol to be adaptive and reliable when some nodes undergo failures. The robustness of GPSR and RPL with PPS against node failures is demonstrated in Fig. 5.21 when network availability a is swept. Note that, in this study, each node is assumed to generate data at the base data rate and shadowing variance σ and cluster size n are constant at 8.0 dB and 1,000 nodes, respectively. The network becomes unstable with the presence of increasing random node failures. Therefore, the transmission reliability is decreased for both GPSR and RPL. For GPSR, when a decreases from 100 to 90 %, P_D slightly

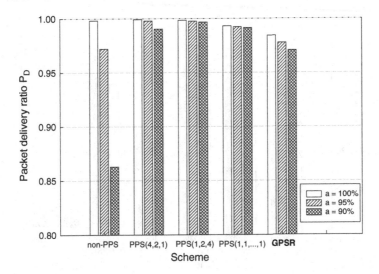

Fig. 5.21 PDR of different routing schemes versus network availabilities (case V)

decreases from 98.43 to 97.07 %. Additionally, accurate neighbor information is maintained when neighborhoods change by an adequate hello message rate. Nodes could promptly correct routing paths and thus packet losses due to unreachable nodes are mitigated. For RPL-non-PPS, although its P_D is 99.82 % when a is 100 %, it drops to 85.84 % when a is 90 %. All transmissions to a primary preferred parent that has already failed are useless, the performance is therefore significantly degraded. In comparison with GPSR and RPL-non-PPS, RPL with PPS also experiences degraded P_D with decreasing a. Fortunately, P_D drops at a much lower rate. This can be explained due to the fact that PPS helps to deflect traffic from failed nodes by switching the next hop to another parent. For example, RPL-PPS(4,2,1) gives the best performance when a is high (i.e, a is 100 % or 95 %). RPL-PPS(1,2,4) gives more retries to parents with higher relative ranks and thus results in higher-cost paths. RPL-PPS(1,1,...,1) is over-reacting (i.e., it reacts too fast to transient network condition changes and results in unnecessary path fluctuations) and at the same time tends to push next hops to parents with higher costs than those of primary preferred parents. As a result, both RPL-PPS(1,2,4) and RPL-PPS(1,1,...,1) experience lower PDRs as compared to RPL-PPS(4,2,1). However, when more nodes become unavailable, packets should be deviated quickly to have a higher chance to reach any active parents. Therefore, when a decreases to 90 %, RPL-PPS(1,2,4) begins to give a higher P_D, i.e., 99.69 %, than that of RPL-PPS(4,2,1), i.e. 99.03 %.

Figure 5.22 plots D_{95} of transmission delays. As can be seen, delays of GPSR and RPL-non-PPS are slightly reduced when network availability decreases. Packets that have experienced significant delays after traversing over long paths with increasingly failing nodes will have a higher likelihood of being lost. Note that the

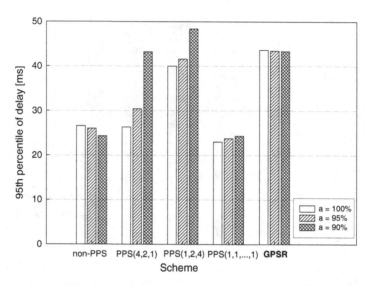

Fig. 5.22 Ninety-fifth percentile of transmission delays of different routing schemes versus network availabilities (case V)

delay values shown in Fig. 5.22 only count delays for successfully delivered packets. Therefore, the decreasing network availability decreases transmission reliability and delays of GPSR and RPL-non-PPS. On the other hand, RPL with PPS improves reliability but induces higher delays. With low network availability, both RPL-PPS(4,2,1) and RPL-PPS(1,2,4) might give too many wasteful attempts to primary or secondary parents that might have already failed. Therefore, when a is 90 %, D_{95} of RPL-PPS(4,2,1) and PPS(1,2,4) are increased to 43.20–48.38 ms, respectively. While RPL-PPS(1,1,...1) has the lowest D_{95}, i.e., 24.34 ms, since it does not perform the exponential back-off procedure for packet re-transmission. Once a transmission failure is detected, nodes will simply re-transmit the packet to another candidate without exponentially increasing the contention window.

Next, in order to have a more detailed understanding of the above results, mean hop count for various schemes are compared in Fig. 5.23 (with an assumed network availability of 90 %). Since GPSR takes effort in forwarding packets to the nodes closest to the destination, it pursues the goal of minimizing hops traversed between the source and the destination. As can be seen in Fig. 5.23, GPSR forwards most packets over short paths to reach the destination. This is demonstrated by having higher fraction of paths with less than 10 hops and no path longer than 16 hops. For RPL-non-PPS, packet transmissions always choose primary preferred parents having the lowest relative ranks, the traffic flow therefore also follows short paths to the root, where the maximum path length is 16 hops. However, in order to maximize the transmission reliability, RPL-non-PPS chooses parents with better link qualities. More reliable links (lower ETX value) with shorter distance are chosen for each hop so that RPL has a higher proportion of longer paths than

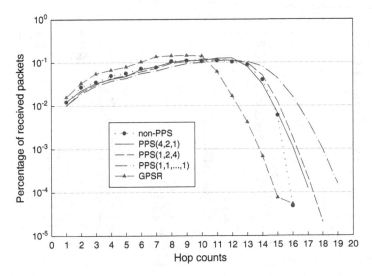

Fig. 5.23 Distribution of routing path lengths versus parent switching schemes ($a = 90\%$, case V)

GPSR, i.e., higher fraction of packets with path length of 12–16 hops. When PPS is employed, routing paths are lengthened since the traffic flow can be re-directed, in cases where preferred parent failures are detected, towards alternate parents having higher relative ranks. In the extreme case, i.e., RPL-PPS(1,1,...,1), there are noticeable fraction of paths that require 17–19 hops due to the fact that, at each node, packets can be spread to 6 different alternate parents (in addition to the preferred parent) that significantly diverge paths from the root. This illustrates the over-reacting behavior and in consequence longer path lengths are observed. RPL-PPS(4,2,1) also selects longer paths (as compared to RPL-non-PPS) to deflect traffic from points of failures, however, over-reaction is prevented by allowing only 2 possible alternate parents. RPL-PPS(1,2,4) has slightly higher fraction of longer paths. Recall that RPL-PPS(4,2,1) prefers the default parents while RPL-PPS(1,2,4) allows more trials with alternate parents. For example, if a node requires five re-transmissions for packet delivery, the former only attempts with two parents whereas the latter will likely involve all three parents. To sum up, by considering more candidate parents when switching, a higher level of path diversity can be obtained. However, increasing path diversity may result in longer paths that may in turn require more network resources and thus degrade network performance.

Mean communication cost of routing paths is plotted in Fig. 5.24. It is measured by the ratio of the total number of frame transmissions in the entire network against the number of packets successfully delivered to the DAP. When the network availability a is 100%, GPSR has higher path costs than RPL since it has to try several re-transmissions over those long but unreliable links in order to have successful deliveries. However, when a decreases, the impact of node failures on the

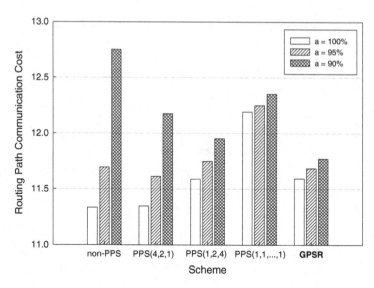

Fig. 5.24 Communication cost versus parent switching schemes and network availabilities (case V)

path cost is negligible. Accurate neighbor connectivity information is maintained and all failed nodes are excluded promptly by frequent hello messages. For a given RPL scheme, as expected, when the network availability (i.e., the number of working nodes) decreases, increasing path costs are observed in Fig. 5.24. This can be explained as follows. A decrease in the number of working nodes might enforce more senders to replace their lower-rank failed parent by higher-rank alternate parents as their next hops. These alternate parents are more likely to require additional transmissions and/or subsequently lead to higher-cost paths to the root. Another observation is that, when there is no node failure, non-PPS scheme works well and results in quite low path costs by sending packets over high quality links connecting the sender to its default parent. However, when nodes failures occur, it may blindly keep sending packets to failed default parents until the DODAG is corrected by conventional local and/or global repairs. Thus, it requires a greater number of transmissions. Compared to RPL-PPS(4,2,1), RPL-PPS(1,2,4) and RPL-PPS(1,1,...,1) require higher routing costs since they tend to deviate traffic to alternate parents that are more likely farther away from the senders and hence lead to links with lower quality and/or longer paths. Note that when network availability a is 90 %, the path cost of RPL-PPS(1,2,4) is slightly lower than that of RPL-PPS(4,2,1). This indicates that when there are significant number of node failures, RPL-PPS(4,2,1) might give too many wasteful attempts to primary parents that have likely already failed. Although RPL-PPS(1,2,4) has the lower path cost than all other RPL schemes, GPSR has the lowest path cost as shown in Fig. 5.24. Nodes update their neighbor status with frequent hello messages. The overall GPSR path cost is relatively lower than that of the PPS mechanism since almost none of the GPSR transmissions are wasted on failed nodes.

All of the above results in this study case evaluate the ability of GPSR and RPL to operate in networks with different levels of network failures. When network availability a decreases, the robustness of RPL in providing reliable communications is enhanced by the PPS mechanism which nominate alternate parents for re-transmissions. However, along with path diversity, transmission delay is increased with the increasing number of re-transmissions via non-optimal and longer paths. Generally, for all considered network availabilities, RPL with PPS significantly improves transmission reliability and still produces tolerable transmission delays.

5.3.2.2   Study Case VI: Effects of Node Failures Under High Load

To study the effects of increasing traffic load in the presence of node failures, the per-node data rate r is swept in this study case. The comparisons of RPL with PPS and GPSR are plotted in Figs. 5.25 and 5.26. Network availability, channel shadowing variance and cluster size are fixed to 90 %, 8.0 dB and 1,000 nodes, respectively. It can be observed that there is a decrease in PDR and an increase in transmission delay for both GPSR and RPL when the traffic load becomes heavier. GPSR achieves lower PDR than RPL with PPS in all cases of different traffic loads. When $r = 0.1$ packets/s/node (more than 50 times the base rate), P_D of GPSR drops to 94.17 % and D_{95} is increased to 57.72 ms. This can be explained by the fact that data traffic intensely compete for channel access and consequently network collisions

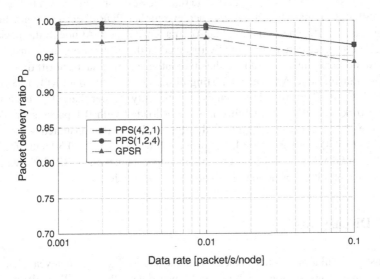

Fig. 5.25 PDR of different routing schemes versus per-meter data rates ($a = 90$ %, case VI)

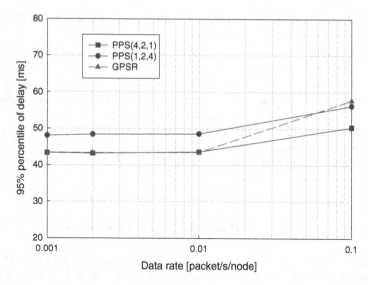

Fig. 5.26 Ninety-fifth percentile of transmission delay of different routing schemes versus per-meter data rates ($a = 90\%$, case VI)

occur. When traffic load is high, GPSR is unable to adapt and thus its performance is further degraded. This problem can be mitigated by dynamic rerouting mechanism of PPS, as indicated in Fig. 5.25. Comparing between GPSR and RPL with PPS, the latter improves the transmission reliability over the entire range of data rates of interest. For illustrative examples, compared to GPSR, RPL-PPS(1,2,4) can boost P_D from 97.07 to 99.69 % at the base data rate. However, as can be seen in Fig. 5.26, RPL-PPS(4,2,1) and RPL-PPS(1,2,4) achieve higher PDRs but experience higher delays than GPSR when r is less than 0.1 packets/s/node. Although the packet is redirected to other parents and finally received at the destination, the transmission delay significantly increases due to the cumulative exponential back-off time before each re-transmission. Another interesting observation is that both RPL-PPS(4,2,1) and RPL-PPS(1,2,4) are able to achieve a reliability greater than 96 % even when the network with 90 % availability is heavily loaded with 0.1 packets/s/node. In this case, the superiority of RPL with PPS is clearly demonstrated where both RPL-PPS(1,2,4) and RPL-PPS(4,2,1) provide lower D_{95}, i.e., 50.39 and 56.23 ms, than that of GPSR, i.e., 57.72 ms. Overall, in comparison with GPSR, RPL with PPS more efficiently handles node failures and heavy traffic load.

5.4 Discussions

Extensive simulation results presented in the preceding section reveal that RPL offers higher transmission reliability and lower delays than GPSR in all scenarios of interest that are characterized by different channel conditions, traffic loads and network sizes. Moreover, under the consideration of multiple node failures, PPS can effectively improve the network resilience since it can adaptively reroute packets over multiple alternative paths. Consequently, the results indicate that RPL with PPS is a very promising routing protocol for NAN communications. However, RPL does impose extra requirements on network signaling overheads, memory and computation capability of network devices since it needs to estimate link quality, maintain neighbor information and the routing tree structure.

References

1. *NIST Priority Action Plan 2 – Guidelines for Assessing Wireless Standards for Smart Grid Applications*, National Institude of Standards and Technology Std., 2011.
2. *Part 11: Wireless LAN Medium Access Control (MAC) and Physical Layer (PHY) specifications*, IEEE Std 802.11b-1999 Std., 2001.
3. T. Watteyne *et al.*, "From MANET to IETF ROLL standardization: A paradigm shift in WSN routing protocols," *IEEE Communications Surveys & Tutorials*, vol. 13, no. 4, pp. 688–707, Jan. 2011.
4. B. Karp and H. Kung, "GPSR: greedy perimeter stateless routing for wireless networks," in *Proc. Annual International Conference on Mobile Computing and Networking (MobiCom)*, Boston, MA, USA, Aug. 2000, pp. 243–254.

Chapter 6
SGCN: Further Aspects and Issues

Previous chapters have provided a big picture, detailed characteristics and requirements of the SGCN, as well as investigations on candidate communications technologies and networking protocols for this network. As an attempt to identify possible future research trends in the SGCN, this chapter outlines a number of technical challenges and corresponding work directions in this network. Specifically, security, machine-to-machine communications, network coding, cloud computing, software-defined networking, network virtualization, smart grids and smart cities are addressed.

6.1 Cyber Security

The SGCN conveys sensor data concerning the energy consumption of millions of customers as well as monitoring/control signals used for grid stabilization and optimization. It is therefore vulnerable to privacy and security issues [1]. For example, smart meters can be attractive targets for hackers largely because they are widely installed in consumer premises and their vulnerabilities can easily be monetized. Hackers who compromise a meter by exploiting software bugs in the exposed infrared port or networking protocols can immediately manipulate their energy costs or fabricate energy meter readings [2, 3]. Widespread smart meter misuse could also have broader effects. Patterns of electricity usage could lead to disclosure of not only customer's energy consumption habits but also their personal activities. Increases in power draw might suggest changes in business operations. Such energy-related information could support criminal targeting of homes or provide business intelligence to competitors. Especially, a successful hack into centralized control systems could take over the control of the power grid in regional or national scales. The consequence is that while the utilities might not be able to operate the grid, malicious hackers, e.g., terrorists, could destroy or fabricate

© The Author(s) 2014   99
Q.-D. Ho et al., *Wireless Communications Networks for the Smart Grid*,
SpringerBriefs in Computer Science, DOI 10.1007/978-3-319-10347-1_6

data, take out potentially hundreds of power plants, substations, transformers, etc. These disruptive actions could result in cascade effects or even blackout of the entire power grid and thus lead to significant losses of technological and financial benefits that the SG is expected to yield as well as a great deal of damage to public safety and homeland security.

One of the challenges in SGCN security is that existing power grid control systems were originally designed for use with proprietary and utility-owned local communications networks. Over the last few decades, to have better system controls at lower costs, utilities have connected them to the public networks without building additional technologies to make them secure. Moreover, most communications technologies and networks currently recommended for use in the SG have not been designed for industrial and critical infrastructures. It is hard for the existing security techniques to meet all security requirements of the SGCN [4,5]. Compared to regular enterprise applications, SG applications require not only a high level of security but also real-time and continuous operations. Therefore, it is necessary to embrace the existing security solutions where they are applicable, such as communications networks within a control center and/or a substation, and develop unique solutions to fill the gaps where traditional enterprise network cyber security solutions are no longer satisfactory [4,5]. From another perspective, a broad national effort is needed for making related regulations and standards, coordinating research and development between academia, industry, and vendor, as well as developing comprehensive recovery strategies for failures [2, 3].

6.2 QoS Differentiation and Provisioning

The SGCN is not only responsible for providing connectivity for a vast number of devices but also meeting the varying QoS requirements of different types of SG applications. Alarm notifications and control signals for critical missions (e.g., fault detection, distribution grid protection and restoration, etc.) require the latency measured in milliseconds and the consequence of failing to deliver such information on time can be catastrophic. Periodic and regular activities (e.g., energy metering, scheduled software/firmware updating, etc.), on the other hand, require reliable communications and can tolerate the latency of a few seconds or minutes [6,7].

Characterizing the performance requirements for various SG applications is important to understanding which communications technologies and applications should be paired [8]. Besides, in order to enable efficient prioritization of certain applications that have critical requirements, the communications network must be able to differentiate and provision different QoS requirements. Furthermore, existing solutions related to QoS differentiation may need to be revisited to cater for SG traffic because the traffic generated by SG applications will likely be quite different from that generated by traditional web browsing, downloading, or streaming applications that are in use today. The SGCN is expected to carry a mix of both real-time and non-real-time traffic generated and distributed across different network segments. Further, the number, type and proportion of traffic classes will affect the complexity and efficiency of any devised resource allocation scheme. Thus, research is needed to allow for a more seamless administration of all anticipated SG traffic types [6, 8]. One of potential solutions could be the RPL [9] using multiple network graphs constructed with different objective functions corresponding to different traffic classes.

6.3 Network Coding

Operations of the SG depend on complex systems of sensors and controllable devices, all of which are tied together through the SGCN. The efficacy and success of emerging SG applications depends on the capability of the SGCN to reliably and expediently gather and transmit sensor data to control centers and vice versa. Transportation of data from a huge number of sensors to a limited number of data collectors or centralized computing centers poses numerous communications challenges due to the bottlenecks associated with the converge-cast nature of this kind of traffic, harsh outdoor environments that lead to severe channel attenuation and shadowing as well as large network scales. With that in mind, conventional network routing can be designed such that the protocols are robust, reliable, self-healing, scalable and possess low overhead. However, there is a limit on what can be achieved through routing alone [10]. A potential solution to enhance reliability, increase throughput and possibly incorporate security would be network coding [10, 11].

As a concept, network coding is a communications paradigm in which the flow of information instead of individual packets is considered. Specifically, intermediate nodes no longer simply store-and-forward packets but transmit a combination of their received packets. This combination, essentially an algebraic function of received packets, is designed such that network performance (i.e., capacity, reliability, security, scalability, data compression, etc.) is improved. To illustrate this concept, consider the basic case where a single link is available but two packets have to be transmitted, transmitting each in turn requires two transmission slots. However, transmitting a combined version of both packets requires only one transmission slot and thus capacity is improved. Of course, to recover both packets, knowledge of the combination function is required beforehand. Beyond capacity concerns, reliability can also be achieved through the use of spatial and temporal diversity. Since each un-coded packet can be disseminated throughout the network over space (over different links) and time (at different times) via multiple coded packets, this diversity allows for enhanced reliability. Moreover, given the encoding operation, in the presence of a limited attacker, i.e., an attacker who cannot overhear all the network edges, a coded packet can be intercepted but not decoded. Thus, a basic form of security can be achieved. With these potential advantages, network coding protocols have been successfully implemented to increase robustness and efficiency in distributed storage systems [12] as well as sensor networks [13].

For the SGCN, network coding's opportunistic use of the wireless broadcast medium and its ability to incorporate diversity in a seamless manner to increase robustness and efficiency of data transmission leads to potential advantages such as data aggregation through coding, data compression through packet correlation, enhanced security, improved reliability as well as scalability. In particular, future research direction is focused on a network coded extension to RPL [14] as well as tunable sparse network coding [15].

6.4   Machine-to-Machine Communications

The intelligence of the SG relies heavily on wide-scale monitoring and control applications for which the intended form of communications is machine-to-machine (M2M). Specifically, M2M communications is characterized by short bursty traffic profiles within a network of various heterogeneous electronic devices, communications and software technologies [16, 17]. Furthermore, given the segmented nature of the SGCN, technology inter-operability is essential to maintain latency and reliability requirements imposed by SG M2M applications. In particular, standardization activities have been underway at the European Telecommunications Standards Institute (ETSI), the 3rd Generation Partnership Project (3GPP), and IETF routing over low power and lossy networks (ROLL). At ETSI, an end-to-end view of M2M standardization is considered. Further, given their mandate by the European Commission on Smart Metering (M/441), they aim to provide access to meter databases, provide end-to-end service capabilities as well as smart metering application profiles. At 3GPP, M2M communications is called machine-type-communications (MTC) and is being examined for the optimization of access and core cellular networks for efficient and secure support of M2M applications. At IETF ROLL, the aim is to extend the IP for sensor technologies and M2M devices (such as building monitoring infrastructures of the SG). Precisely, the focus is on extending RPL with an end-to-end IP-based solution to mitigate inter-operability concerns.

Focusing on the SG, different environments translate into differing challenges and M2M requirements. At the home and smart building level, wireless networks are the norm and applications such as lighting control, heating, ventilation, etc., require SG monitoring applications. Therefore, the main M2M requirements are very low power consumption and inter-operability, coordination and resource allocation between the various different interconnected devices. At the power distribution level, the SG offers cost-effective communications for monitoring, control and diagnostic purposes. At this level, the main requirements are high scalability, high reliability and robustness to harsh power system environments. Furthermore, when considering communication technologies, promising candidate system architectures incorporate cellular networks at the WAN level. This in turn raises further challenges associated with resource scheduling and allocation at the cellular level. Given these challenges, the NAN/WAN interconnection becomes a crucial design feature for achieving an efficient and reliable SGCN. In particular, the resource allocation complexity introduced by the sheer number of M2M endpoints within the NAN must be reduced, the scheduling algorithm complexities must be reduced through either group-based or time granularity scheduling [18] and the wasted capacity generated by cellular control channel bottlenecks must be mitigated by some form of data aggregation [19]. Given these challenges and requirements, a wide range of research areas related to M2M communications in the SG need to be addressed. They include gateway design, M2M computing platform, M2M middleware and application programming interfaces (APIs), service differentiation, security, self-configuration and self-organization.

6.5 Cloud Computing

Cloud computing involves computing over a network, where a program or application may run on multiple connected servers at the same time. It enables the sharing of resources to achieve coherence and economies of scale. In the power grid, rather than having their own servers to run applications, the utilities can move their data processing, computation, and many other applications/services to the cloud. By doing that, they can migrate from the traditional capital expenditures (CAPEX) model (i.e, buying the dedicated hardware and software and depreciating them over a period of time) to the operational expenditure (OPEX) model (i.e., using a shared cloud infrastructure and pay as they use it). Cloud computing allows utilities to avoid upfront infrastructure costs, scale the computing and storage capacities up or down according to their needs, and get their applications up and running faster with improved manageability and less maintenance. Since cloud service providers are responsible for the environment, utilities do not need to maintain a complex computer hardware/software infrastructure and thus can focus on other critical tasks directly related to energy generation and delivery [20, 21]. In addition to financial benefits, cloud computing can efficiently facilitate distributed computing model that would spread control functions of the SG across multiple machines and locations. The advantage would be that if one element of the computing system gets compromised due to hardware/software failures or cyber attacks, the other virtual machines could step in to protect the system and coordinate their efforts to keep the grid functioning properly and smoothly.

While cloud computing model may offer various values and advantages as just mentioned above, a number of challenges must be addressed in order to meet unique requirements of the SG. First of all, the SG handles enormous amounts of real-time data from heterogeneous sources like metering infrastructure, phasor measurement units, feeder load data, maps, cameras, electric vehicles, etc. A cloud-based platform for the SG is thus required to be scalable and consistent when dealing with such kinds of data. It must be able to receive and process a huge volume of data from multiple devices at the same time and make appropriate actionable responses in a timely manner. Next, different from cloud-based platforms providing personal or conventional business services, the one for the SG demands a very high availability. Almost all SG applications require industry-grade connectivity. This in turn requires cloud computing services to be highly available and accessible without interruption. Downtime or data loss may compromise power system operations and thus impose substantial costs [22]. Another issue is that the externalized aspect of cloud-based solutions demands service providers to ensure data privacy and integrity, especially when cyber security is paramount in the SG. Cloud-based platforms for the SG should address the fundamentals of security and risk management through a comprehensive approach placing a strong emphasis on isolation, identity and compliance [22, 23].

There are two basic cloud computing models, i.e., public or private. The public model has advantages in costs thanks to its high level of resource sharing and utilization, However, it has challenges in availability and security due to its public and multi-tenant environment. On the other hand, the private model requires higher costs since the cloud infrastructure is owned and operated by a given organization and provides services within its own perimeter. This model is, fortunately, able to provide higher levels of availability and security [22]. The utilities may employ a hybrid cloud that integrates both public and private cloud models. The public SG cloud is provided by public vendors for non-critical applications such as geospatial imaging, mapping, and weather forecasting. The private SG cloud is developed and owned by government organizations (e.g., the department of energy) and to be used primarily for energy data management, reporting, analytics, command and control functionalities. As a result, in addition to fundamental technical issues related to availability, security, and time responsiveness, a variety of national and local policies, management frameworks, and standards need to be developed for a successful implementation of SG cloud computing platform. This calls for significant collaborative work among the government, standards development organizations, utilities, vendors, and service providers [22, 24].

6.6 Software-Defined Networking and Network Virtualization

As presented in the previous section, the sharing and networking of computing resources enabled by cloud computing technology are promising for an economical and flexible implementation of SG applications and services. More recently, the paradigm shift in the service model for server resources has been extended to the network infrastructure with the introduction of software-defined networking (SDN) and network virtualization.

SDN is a new approach to have a better communications network design, implementation and management. In a nut shell, it separates the control and forwarding planes of the network to make it easier to optimize them. Specifically, a centralized controller provides an abstract and global view of the overall network.

Through the controller, network administrators can quickly and easily make and push out decisions on how the underlying systems (e.g., switches and routers) of the forwarding plane will handle the traffic [25]. The most common protocol used in SDN networks to facilitate the communications between the controller and the underlying systems is the OpenFlow [26]. Another key feature of SDN is that it provides open APIs to support all services and applications running over the network. These APIs facilitate innovation and enable efficient service orchestration and automation. As a result, SDN enables a network administrator to shape traffic and deploy services to address changing needs and requirements without having to touch each individual switch or router in the forwarding plane [27].

Virtualization is a concept that allows the abstraction, sharing and partitioning of resources. The goal of virtualization is to enable a more dynamic and efficient reuse of resources. In the so-called server or computer virtualization, computing resources such as processors, memory and storage are shared. For example, a virtual machine (VM) is a fully-functional instance of the physical hardware. For network virtualization, the network switching fabric and backplane is shared. In the case of wireless virtualization, the wireless resources such as access points, base stations, and network interfaces are shared. SDN is considered as a facility for the implementation of network virtualization [28].

The SDN and network virtualization can be adopted to develop and deploy a centralized, programmable, and evolvable SGCN in order to reduce CAPEX and OPEX, enhance network performance and flexibility, and so on. Efficient dynamic resource allocation and traffic routing in the SGCN can be implemented with SDN based on global knowledge of the overall network and on-demand reconfigurability of switches, routers, etc. Network virtualization allows the utilities to offer fully-differentiated services to different types of SG traffic by having total control over their own virtual network infrastructure. It also enables the hardware equipment to be reprogrammed instead of specialized. This feature is especially useful in the case of the SGCN where the utilities desire to control and manage the telecommunication network in order to meet the QoS requirements of mission-critical applications and services.

However, since SDN and network virtualization are still in their development stages, there are a number of key challenges to overcome before they can be utilized in the SGCN. For example, the centralized nature of the controllers coupled with an increasing in number of access points and devices installed at customer premises over which the grid operators have no direct physical control, can potentially make it vulnerable to cyber attacks. Distributed denial-of-service (DDoS) attacks launched from compromised smart meters and appliances can compromise the operations of the network. Besides, guaranteeing the confidentiality, integrity, and availability of control signals as well as SG information flow is essential. Finally, the SGCN, as a critical utility infrastructure, is expected to meet carrier-grade requirements including scalability, reliability, QoS, etc. The current OpenFlow standard for SDN provides only limited QoS support and has difficulty achieving fast failure recovery that is necessary for high reliability due to its dependency on a centralized controller [29, 30].

6.7   Smart Grids and Smart Cities

The electric power grid is the most important feature in any city. If the electric energy is unavailable even for a short period of time, nearly all functions of the city will be affected or stop working and thus businesses, industries, social activities, public safety and security, and so on will be shaken. Recently, the smart grid has been playing a decisive role in the realization of smart cities. A smart city encompasses numerous intelligent city operations and services related to transportation, parking, housing, water, waste, education, healthcare, safety, security, etc., in order to improve the economic, social, and environmental health of the city. The smart grid essentially provides an advanced energy infrastructure that links together different elements of these operations and services, supports various techniques and applications to improve their efficiencies, and, most importantly, enables coordination between city staff, infrastructure operators, urban planners, and citizens [31]. Conversely, smart city initiatives can help build consumer awareness of energy efficiency initiatives. From the utilities' perspective, tying smart meter programs into a broader smart city project enables a closer engagement with communities, individual consumers, and businesses [32]. Besides, transportation and traffic systems would coordinate with the energy systems to support critical transportation arteries and modes. Through it all, timely logistics information would be gathered and supplied to the public through social media networks and other public communication means. Conservation, efficiency and safety will all be greatly enhanced through the availability of accurate logistical information [31].

While smart grids and smart cities are strongly correlated and their integration is highly desired, there are still many challenges. The biggest challenge is that even though there have been many pilots developing and implementing smart grids together with smart cities, it is still difficult to extend them into city-wide deployments. Large-scale deployments need to be tied in with utilities' plans that are largely determined by financial budgets, user acceptance levels, industry regulators, national government policies, etc. This means that the interests of the smart city might not be the primary concern to the utilities. For this reason, most large-scale smart meter deployments are being considered in parallel with smart city projects with only limited integration. City authorities and utilities need to work closely together to ensure that they do not lose the benefits of an integrated approach, particularly with regard to consumer engagement. Cities also need to develop a master energy plan if they are to ensure that the various elements of their energy strategy are fully integrated. Failure to take a coordinated approach could leave a legacy of isolated pilot projects and an inadequate energy infrastructure for the needs of a smart city [33].

Besides, to facilitate the integration of smart grids and smart cities, the utilities may need to share their information and communications infrastructure with other organizations or make use of that of other organizations. For example, information on power failures or outages detected by the utilities may need to be shared with corresponding local police stations and healthcare institutions for public security and safety purposes. The utilities may exploit information on consumer behaviors, social trends, urban planning from the city to develop their power generation and distribution strategies. As a result, great coordination efforts in making policies and regulations, as well as developing relevant communications technologies and interoperability standards are necessary.

Finally, electric vehicles (EVs) and autonomous (or self-driving) vehicles (AVs) are emerging as revolutionary automotive technologies that involve both smart grids and smart cities. More than 35 million EVs will be on roads worldwide by 2022 [34]. Sales of AVs will grow from fewer than 8,000 annually in 2020 to 95.4 million in 2035 [35]. These two technological innovations may drive many smart grid and smart city technologies and their deployment road maps. Smart EV charging infrastructure is to be constructed around the city. It is responsible to charge vehicles in a cost-efficient manner while ensuring consumer's comfortability. Vehicle-to-grid (V2G) infrastructure allows EVs to communicate with the power grid to sell demand response services by either delivering electricity into the grid or by throttling their charging rate. Smart city technologies such as global system positioning, coordinated traffic signaling, smart parking, and street lighting are needed for the arrival of AVs. These technologies need to be developed jointly to exploit their full potentials and pave the road for applications and services to be emerged in future.

References

1. F. Cohen, "The smarter grid," *IEEE Security & Privacy*, vol. 8, no. 1, pp. 60–63, Jan. 2010.
2. P. McDaniel and S. McLaughlin, "Security and privacy challenges in the smart grid," *IEEE Security & Privacy*, vol. 7, no. 3, pp. 77–87, May 2009.
3. K. Khurana *et al.*, "Smart-grid security issues," *IEEE Security & Privacy*, vol. 8, no. 1, pp. 81–85, Jan. 2010.
4. "Securing the smart grid," Cisco Systems, white paper, 2009.
5. Y. Yan *et al.*, "A survey on cyber security for smart grid communications," *IEEE Communications Surveys & Tutorials*, vol. 99, pp. 1–13, Jan. 2012.
6. K. Hopkinson *et al.*, "Quality-of-service considerations in utility communication networks," *IEEE Transactions on Power Delivery*, vol. 24, no. 3, pp. 1465–1474, Jul. 2009.
7. S. K. Tan *et al.*, "M2M communications in the smart grid: Applications, standards, enabling technologies, and research challenges," *International Journal of Digital Multimedia Broadcasting*, pp. 1–8, 2011.
8. G. W. Arnold, "Challenges and opportunities in smart grid: A position article," *Proceedings of the IEEE*, vol. 99, no. 6, pp. 922–927, Jun. 2011.
9. "RPL: IPv6 routing protocol for low-power and lossy networks," Internet Engineering Task Force (IETF), RFC 6550, March 2012.
10. A. S. M. Medard, *Network Coding: Fundamentals and Applications.* Academic Press, Oct. 2013.
11. R. Bassoli, H. Marques, J. Rodriguez, K. Shum, and R. Tafazolli, "Network coding theory: A survey," *IEEE Communications Surveys & Tutorials*, vol. 15, no. 4, pp. 1950–1978, Apr. 2013.
12. A. Dimakis, P. Godfrey, Y. Wu, M. Wainwright, and K. Ramchandran, "Network coding for distributed storage systems," *IEEE Trans. Information Theory*, vol. 56.

13. J. F. A. Kamra, V. Misra and D. Rubenstein, "Growth codes: maximizing sensor network data persistence," in *Proc. the 2006 conference on Applications, technologies, architectures, and protocols for computer communications (SIGCOMM)*, Oct. 2006, pp. 255–266.

14. "Network coding for enhancing data robustness in low-power and lossy networks," Network Working Group, Internet Draft, Nov. 2013.

15. R. Prior, D. E. Lucani, Y. Phulpin, M. Nistor, and J. Barros, "Network coding protocols for smart grid communications," *IEEE Transactions on Smart Grid*, vol. 5, no. 3, pp. 1523–1531, May 2014.

16. Z. Fadlullah, M. Fouda, N. Kato, A. Takeuchi, N. Iwasaki, and Y. Nozaki, "Toward intelligent machine-to-machine communications in smart grid," *IEEE Communications Magazine*, vol. 49, no. 4, pp. 60–65, Apr. 2011.

17. S. K. Tan, M. Sooriyabandara, and Z. Fan, "M2M communications in the smart grid: Applications, standards, enabling technologies, and research challenges," *International Journal of Digital Multimedia Broadcasting*, vol. 2011, 2011.

18. Y. Xu and C. Fischione, "Real-time scheduling in lte for smart grids," in *Proc. the 5th International Symposium on Communications Control and Signal Processing (ISCCSP)*, 2012, pp. 1–6.

19. J. Brown and J. Y. Khan, "Key performance aspects of an LTE FDD based smart grid communications network," *Computer Communications*, vol. 36, no. 5, pp. 551–561, 2013.

20. T. Singh and P. K. Vara, "Smart metering the clouds," in *Proc. the 18th IEEE International Workshops on Enabling Technologies: Infrastructures for Collaborative Enterprises (WET ICE)*, Jun. 2009, pp. 66–71.

21. A. Mohsenian-Rad and A. Leon-Garcia, "Coordination of cloud computing and smart power grids," in *Proc. the first IEEE International Conference on Smart Grid Communications (SmartGridComm)*, Oct. 2010, pp. 368–372.

22. "Security and high availability in cloud computing environments," IBM Global Technology Services, white paper, Jun. 2011.

23. "Cloud security guidance - IBM recommendations for the implementation of cloud security," IBM, technical paper, 2009.

24. Y. Simmhan *et al.*, "Cloud-based software platform for data-driven smart grid management," in *Proc. IEEE/AIP Computing in Science and Engineering*, 2013.

25. S. Paul, J. Pan, and R. Jain, "Architectures for the future networks and the next generation Internet: A survey,," *Computer Communications*, vol. 34, pp. 2–42, Jan. 2011.

26. N. McKeown *et al.*, "OpenFlow: Enabling innovation in campus networks," *SIGCOMM Computer Communication Review*, vol. 38, 2008.

27. "Network functions virtualisation: An introduction, benefits, enablers, challenges & call for action," SDN and OpenFlow World Congress, white paper, Oct. 2012.

28. A. Wang, M. Iyer, R. Dutta, G. N. Rouskas, and I. Baldine, "Network virtualization: Technologies, perspectives, and frontiers," *Journal of Lightwave Technology*, vol. 31, no. 4, pp. 523–537, Feb 2013.

29. J. Zhang, B.-C. Seet, T.-T. Lie, and C. H. Foh, "Opportunities for software-defined networking in smart grid," in *Proc. the 9th International Conference on Information, Communications and Signal Processing (ICICS)*, Dec. 2013, pp. 1–5.

30. "OpenFlow-enabled SDN and network functions virtualization," Open Networking Foundation, solution brief, Feb. 2014.

31. K. Geisler, "The relationship between smart grids and smart cities," IEEE Smart Grid, Tech. Rep., May 2013.

32. E. Woods, "Why smart cities need smart grids," Navigant Research, Tech. Rep., 2013.

33. "Why Europe needs smart grids and smart cities," Smart Utilities, Scandinavia. [Online]. Available: http://www.smartutilitiesscandinavia.com.

34. S. Shepard and J. Gartner, "Electric vehicle market forecasts," Navigant Research, Tech. Rep., 2013.

35. D. Alexander and J. Gartner, "Autonomous vehicles," Navigant Research, Tech. Rep., 2013.